Truss Fun

Second Edition

by

David W. Harris, PhD, PE

Published by:
BaHa Enterprises, LLC
PO Box 261173
Lakewood, CO 80226

Fungineering
We put the fun in engineering

Truss Fun

Published by:
BaHa Enterprises, LLC
PO Box 261173
Lakewood, CO 80226

Printed in the United States
By Whitehall Printing Company
Naples, Florida

Library of Congress Catalog Control Number
First Edition: 99 097177
Second Edition: 2003099306

Disclaimer

This book is intended for educational purposes. Historical references are used for information. General principles and example material property data are used to explain engineering. Design and construction of real structures should only be accomplished by a registered engineer. This book is not intended as a reference for design purposes.

Liability Release

BaHa Enterprises, LLC is not responsible in any way for results from attempting projects, activities, or other material contained in this book. ALWAYS BE SAFE. We are building newspaper bridges and other projects that will probably fail with a big enough load. Be careful of falling hazards, and other safety issues. This project should be treated like gymnastics, or other events, where falling is a possibility. Be careful of glue and other materials, follow all manufacturer's warnings and instructions.

On the Cover:
Photo: Courtesy Colorado Historical Society:
Denver & Rio Grande RR Bridges
Gunnison River near Grand Junction, CO
Drawing: Courtesy John Maxwell Collection
Explanation and credit: Chapter 4

Preface

The American Society of Civil Engineers has categorized engineering as a stealth profession, that is work which most people don't know about or understand. Yet, engineering is around us every day in things that we see and the way things work. I hope to provide an interesting and fun way for everyone to experiment and understand some basic concepts of engineering. Engineering isn't stealth, it is science and common sense combined.

Imagine designing something as important as a railroad bridge for the first time over 100 years ago. With a pencil, a big piece of paper, maybe a slide rule, and some basic principles, the engineers set out to conquer the gorges, rivers, and passes with truss bridges. We still use these principles today. These turn of the century examples are used in this book for fun and interest. Big problems were solved without computers or even calculators. Walk through time and through easy principles to learn about engineering, make a hobby more fun, or just to understand the history of science. Most important - make it fun.

Since writing the first edition I have met many great students. They were smart, they were enthusiastic, they were fun, they were good engineers. In the second edition some of their work is shown. Try the challenge, see if you can build a bridge from rolled newspaper and walk across!!!

The Wynkoop Street Railroad Bridge, Denver CO
Excerpt from Historic Plaque, Lower Downtown Historic District

The Wynkoop Street bridge was used by the Denver Rio Grande Railroad as a mainline to supply industry and warehouses. This steel version from 1908, now in use as a pedestrian overpass, serves as a reminder of a once frantic past.

Denver's first settlement was along the confluence of the South Platte River and Cherry Creek. Gold miners and pioneers poured into the area in the late 1800's and were shortly followed by the railroads to supply the community. The first railroad was the Denver Pacific, later part of the Union Pacific, and today a part of the Burlington Northern. Cherry Creek has seen 7 railroad bridges since 1887, 5 spans still stand. After the 1885 flood, bridges were constructed with iron and steel.

Acknowledgments

This book would not be possible without some key help over the years and along the way. My wife, Terry, has always encouraged me to write a book and has been key in advising and editing in the final production of this book.

Early leads on historical references were provided by Gordon Davids of the Federal Railroad Administration. Ron Kaye from the Illinois Central provided key insight into truss projects in the railroad industry and review and suggestions of the manuscript. The Travers family, Fred, Sue and Ben, reviewed and edited the manuscript for educational uses. Our neighbor and friend Jamie Schubert made extensive comments for readability. Caroline Mohorovic, a student and friend, also provided editing of the manuscript.

Historical background information was obtained from a number of sources. The staff and services of the Library of Congress provided important historical work for the chapter on real bridges. The Association of American Railroads (AAR) staff provided information, answered questions, and allowed material from their work to be published. Peter Conlon from AAR gave insights into the railroad industry and helped with leads for materials. Bruce Maxwell from the John Maxwell Collection and the Colorado Historical Society allowed copyrighted information to be included. The American Institute of Steel Construction and Plastruct Inc. allowed reference values to be tabulated for comparison. Finally, the Ontario and Western Railway Historical Society, Inc (O&WHS) provided time, advice, and effort to the production of this book; and have had the vision to provide research facilities for railroad interest and to the nation's history. Art Robb, the O&WHS archivist helped to find and supply some of the historical background. Some of the excellent work from the O&WHS is in Chapter 4.

The Colorado MESA organization and its network of advisors throughout Colorado have been key in helping sponsor activities for High School students, based on the principles of this book. The students' work is shown in the new section on Newspaper Bridges. Thanks to Gloria Nelson, Executive Director for Colorado, Dora Gonzales, Director in Colorado Springs and all the great teachers for their support and passion for teaching and helping young people.

Coos Bay Bridge (Conde B. McCullough Memorial bridge), North Bend, Oregon.
(1934 Public Works Administration cantilever truss bridge)
Photo reproduced from Historic American Buildings Survey/Historic American Record
Library of Congress

Table of Contents

Chapter 1 Introduction 9

Chapter 2 Types of Trusses 11
- 2.1 Definitions for Trusses 11
- 2.2 Deck and Through Trusses 12
- 2.3 Wood Trusses 12
- 2.4 Metal Trusses 13
- 2.5 Joint Construction, Pins and Gussets 13

Chapter 3 Basic Principles of Trusses 17
- 3.1 The Triangle Building Block 17
- 3.2 External loads on trusses 18

Chapter 4 Some Real Bridges 21
- 4.1 Bollman Pinned Truss 21
- 4.2 Warren Truss 26
- 4.3 Howe Truss 32
- 4.4 Pratt Truss 35
- 4.5 Waddell A Truss 37
- 4.6 Lyons Brook Truss and Plate Girder Bridge 41
- 4.7 Bridge Supports - Trestles and Piers 47

Chapter 5 Geometric Scale - Making models 49
- 5.1 Calculating a scale 49
- 5.2 Scales from a ruler called a scale 49
- 5.3 Standard sizes and shapes of construction materials 52
- 5.4 Design properties of construction sections 52
- 5.5 Hobby Modeler's Design Tables 53
- 5.6 Making a HO Model 58

Chapter 6 Analyzing Trusses 61
- 6.1 The Directional Cosines 61
- 6.2 Vectors 61
- 6.3 Static Equilibrium 62
- 6.4 Boundary Conditions 62
- 6.5 Free body diagrams 63
- 6.6 Determinant Trusses - Methods of Pins and Sections 66
- 6.7 Indeterminant Trusses - Springs and Finite Element Method 66
- 6.8 Computer Program TF2DFE 69

Chapter 7 Strength of Materials 79
- 7.1 Properties of Linear and Elastic 79
- 7.2 Principles of stress and strain 80
- 7.3 Stress-strain diagrams and Hooke's Law 81
- 7.4 More Material Characteristics 82
- 7.5 Stresses in Bending 82

Chapter 8 Structural Stability 85
8.1 Conditional Stability 85
8.2 Failure of joints and connections 85
8.3 Buckling 86
8.4 Foundations, Floods, Landslides and Earthquakes 86
8.5 Impacts from Trains 86
8.6 Selected Historical cases 86

Chapter 9 Modeling Failures 89
9.1 Laws of Similitude 89
9.2 Strengths of modeling materials 93
9.3 Time scaling 96

Chapter 10 - Summary 97

References 99

Contributors to Text 100

Workbook: Fun For Students - Projects and Problems 101
Chapter 1 List exercise 102
Chapter 2 Drawing some trusses 103
Chapter 3 Project: Quick Trusses 104
Additional Information: Metric Units 105
Chapter 4 - Historical Research 106
Advanced project 106
Chapter 5 Making Models Practice Problems 107
Advanced project 107
Sources of Additional Information 107
Chapter 6 Analysis of Trusses 108
Advanced Project 107
Additional information: Mathematical Aspects of the Finite Element Method 109
Chapter 7 Strength of Materials Practice Problems 113
Testing Materials at Home 113
Chapter 8 Project: Testing Structures at Home 115
Chapter 9 Materials Modeling 116
Advanced project 116
A Student Project 116
Terms used in this book: 126
Certificate of Achievement. 127
Answers to selected Problems 129
Table of Historic Bridges by State 133
The Newspaper Bridges 143
Types of Bridges. 143
The Newspaper Suspension Project 147
The Newspaper Truss Project 153
The West Point Bridge Designer and Contest. 156
Use of Geometry to Resolve Internal Forces 157
Index 159

Chapter 1 - Introduction

Somehow it seems only natural that when we encounter a small stream, we try to jump over it. When encountering a wider stream, we see if we can toss a stone all the way across. The desire to cross and see what is on the other side is somewhat inherent in human nature.

The desire by people to travel about was one of the earliest incentives to build bridges. Trade and transport of goods were soon to follow. Early bridges date to the days before Christ. Early bridge builders used large flat rocks laid on stones in the river and trees felled singly or together. During the Roman era developments were made in bridge building techniques using stone and an early form of concrete to make circular arches. Simple trusses were constructed in this period. Made of wood, these trusses could span 20-60 feet. These simple designs are still common today.

The advent of the railroad brought a sudden demand for bridges. Developments in bridge design and construction progressed simultaneously in Europe and in the United States. In Europe, the main efforts were focused on the development of the metal girder and design of the arch. In the United States, however, interest focused on the development of the truss, first in wood, later in iron, and finally in steel. It was only natural that with the wealth of timber the United States would develop principles suitable to design with wood.

The truss is one of the major types of engineering structures today. Generally, the truss is made by forming triangles with members which are joined at the intersections with a connection (a joint). Loads applied to the truss are carried along members.

Figure 1.1 - Gunpowder Bridge, Maryland, 1892
Library of Congress, Prints and Photographs Division, Detroit Publishing Company

Trusses are used in many applications. They can be spotted when house roofs are being constructed, in towers that support power lines, as well as in bridges. Trusses may be small or they can be large.

Simple, small trusses were used initially. As longer spans were needed, more complicated trusses came into use. What makes simple trusses fun is that they can be understood using simple principles of math and engineering. This book explains these simple principles in a fun way.

These principles are the foundation of Civil Engineering and other fields of engineering. Much of engineering can be intuitive as it is the study of the constructed features around us all of the time.

In the following chapters examples and drawings are used to help visualize the principles discussed. Bridges are used to make the explanations but watch for trusses all around you.

Figure 1.2 A Power Tower

Figure 1.3 A Pedestrian Overpass
Rocky Mountain Foothills in background

Chapter 2 - Types of Trusses

Beginning in the middle ages the earliest trusses were constructed of wood. The need for bridges associated with the construction of railroads became the greatest factor in bridge design in the United States in the 1800 to early 1900 period. Crossings of large rivers and wide valleys made longer bridges a necessity. The designs of trusses would lead to different types of trusses as this development took place.

2.1 Definitions for Trusses

A typical truss is shown in Figure 2.1 with parts labeled which have common names. A truss bridge usually consists of two trusses which both span the river or valley and are set side-by-side. Members, known as cross members, hold each truss apart and upright. Traffic may pass between the trusses on the cross members, referred to as a through truss, or may run on the upper chord, referred to as a deck truss, see next section for details.

A support is the location where a truss sits on the ground, or other support such as a tower from the river. The span of a bridge is the distance between supports.

A joint is the intersection of multiple pieces of the truss. Vertical members in a truss have specialized names based on their loading condition. Hangers are vertical members in tension and posts are vertical members in compression.

Members are parts of the truss between joints. Verticals are members positioned vertically up and down. Diagonals are members that run from joint to joint and are not vertical. The lower chord is all members across the bottom of the truss; the upper chord is all members across the top of the truss. A panel is all members between two joints in the upper and lower chord, e.g. in Figure 2.1, there would be two triangular end panels on each end of the truss, and rectangular panels across the remainder of the truss.

Each member in a truss carries a force which is assumed to act only along the length, or normal to the member. Forces which push toward a member, or cause the member to shorten are compression forces; forces which pull away from a member, or cause a member to lengthen are tension forces. (Forces are covered in more detail in Chapter 3).

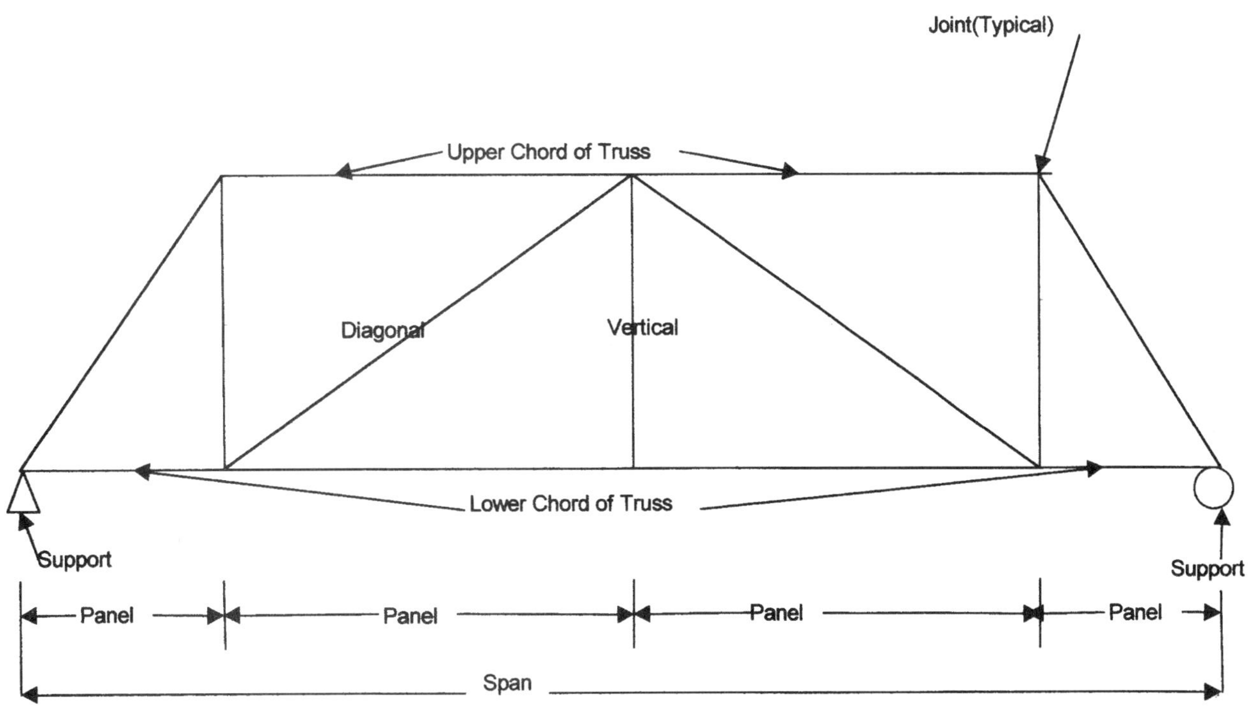

Figure 2.1 Definitions of selected Truss elements

2.2 Deck and Through Trusses

The loads and location for a bridge determine the use of a deck or through truss. The through truss which carries traffic on the lower chord and between the trusses provides greater clearance under the bridge. Cross members, or floor members, must be heavy to carry the weight of traffic and transfer this weight to vertical hangers. As the two supporting trusses must be wide enough to allow traffic to pass through, the truss is generally wider. A deck truss carries load on the upper chord which allows closer positioning of the truss members to carry the load. In many cases this allows load to be placed directly on the truss, eliminating large load carrying cross members. This requires less material and generally is the cheaper between bridge options.

2.3 Wood Trusses

The earliest known trusses were constructed of wood in the Middle Ages. These simple bridges are known as the King Post Truss.

Figure 2.2 King Post Truss

The span varies from 20-60 feet. A variation of the King Post Truss is the Queen Post Truss, a lengthened version of the King Post Truss, useful for spans of 20 to 80 feet. Both truss types are still in use today.

Figure 2.3 Queen Post Truss

When trusses were required for longer spans, the use of the King Post Truss and Queen Post Truss became impractical. New designs emerged which would utilize a repeating pattern of panels across the truss.

Ithiel Town patented the Lattice Truss in 1820. The top and bottom chords held diagonal planks together with the use of spikes created from wooden pins. An advantage of this design was that conventional lumber was used in the construction.

Figure 2.4 - The Town Lattice Truss

In 1840, William Howe of Spencer, Massachusetts patented a wood truss which bears his name, the Howe Truss. Perhaps, the most famous of wood truss types, this design has been used from its invention into the 20th century, and has been the basis for many other designs. The original design was of wood diagonals with verticals of metal; diagonals carried compression loads and verticals carried tension loads. This bridge type was highly favored due to its ease of construction and low cost. The length of Howe Trusses varies from short spans to 150 feet. Howe Trusses would also be constructed of metals as metalwork emerged towards the end of the century.

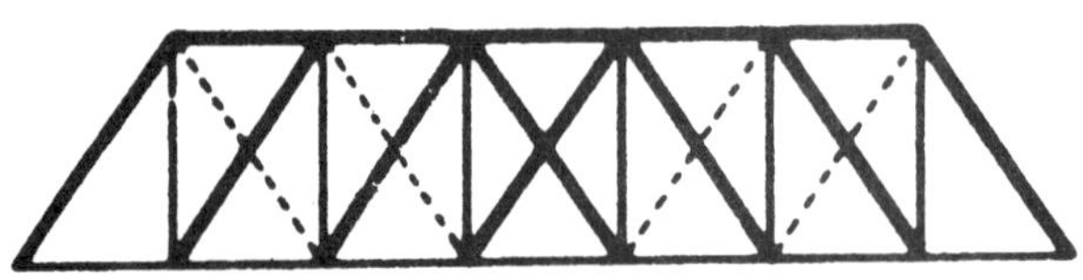

Figure 2.5 - A Howe Truss

2.4 Metal Trusses

Some early metal trusses were developed by Wendall Bollman (1852) and Albert Fink (1854) of the Baltimore and Ohio Railroad. A Bollman truss with a span of 124 feet was built at Harper's Ferry in 1852. An example of a Bollman Truss is shown in Chapter 4. In the same year, a Fink bridge of 3 spans of 205 feet each was constructed over the Monongahola River. With the use of iron and later steel, other designs would emerge.

The Pratt Truss was designed in 1844. This design placed diagonals in tension and verticals in compression. The Pennsylvania Railroad promoted this type of truss from 1851 to 1861. This truss type is still in use today; spans can be up to 250 feet in length.

Figure 2.5 - The Pratt Truss

Warren Trusses were patented in 1848 by James Warren and Theobald Manzani of England. The true Warren Truss is constructed with equilateral triangles. Other Warren Trusses are used which always show triangular outlines. The diagonals carry both compression and tension forces. Warren Trusses generally can span distances of 50 to 400 feet.

Figure 2.6 Warren Truss

Most truss designs are variations of these initial versions. From the Pratt Truss would come variations named the Lenticular Truss, the Parker Truss, the Baltimore Truss, and other new designs. The Warren Truss would be modified and called the Warren with verticals.

As the speeds and weights of trains increased, the stiffness of trusses became increasingly important. The Pratt Truss and Warren Truss were well suited to this need; particularly the Warren Truss which was well suited for riveted connections. In the 20th century, the K Truss was developed, so called because of the distinctive outline of the structural members. Spans of K Trusses vary from 200 to 800 feet. As other construction methods and materials were also being introduced in this time period, few K Trusses have actually been built.

Eighteen types of truss bridges are shown in Figure 2.9 at the end of the chapter. These examples were chosen to show simple trusses, trusses constructed primarily of wood, trusses based on rod and cable braced diagonals, and trusses constructed with steel.

2.5 Joint Construction, Pins and Gussets

Early trusses were pinned at joints, the joint was literally connected by a single round bolt, rivet, or round pin of some type, see Figure 2.7. The forces with a pinned joint are readily found by simple mathematic formulas (see Chapter 6).

Later joints would incorporate large plates, known as gusset plates, at a joint which allowed members to be fastened into the truss without all ends coming together to be fastened by a single pin, see Figure 2.8. The gusset plate, allowed much easier erection of trusses. For analysis purposes (Chapter 6), the truss designs are more complex, leading to indeterminant trusses that require more sophisticated analysis techniques.

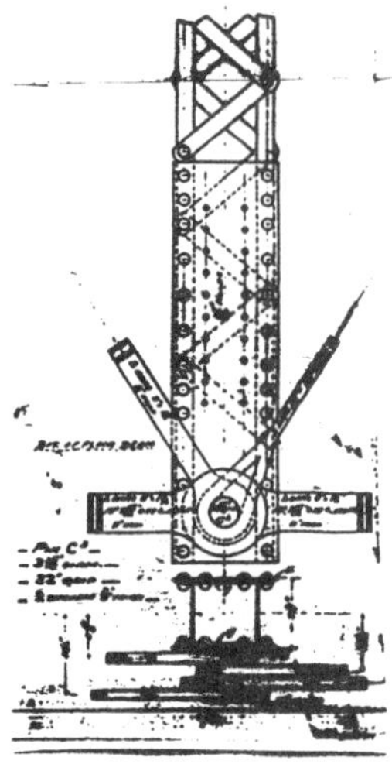

Figure 2.7 Actual Construction Pin Detail
Photo: Reproduced from Historic American Buildings Survey/ Historic American Record On-Line Collection, Library of Congress
Drawing: Denver Rio Grande Railroad
Reprinted with permission: John Maxwell Collection, 1055 Fernley, Jessica NV 89408

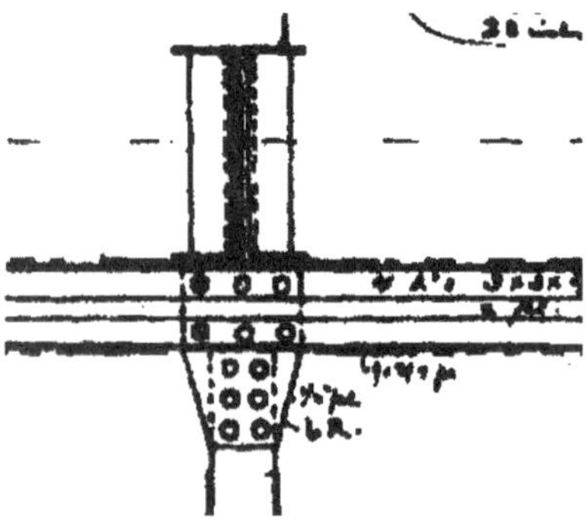

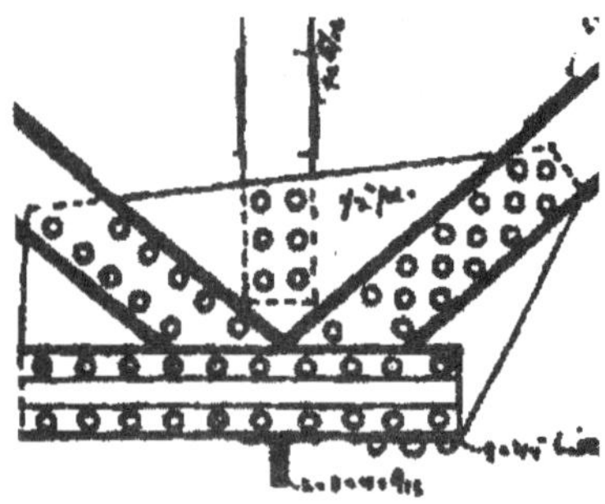

Figure 2.8 Examples of Gusset Plates
Lyons Brook Bridge
Reprinted with permission:
Ontario & Western Historical Railway Society

King Post
(Wood)
Origin in Middle Ages

Length 20-60 feet

Queen Post
(Wood)
Lengthened version of
King Post
Length 20-60 feet

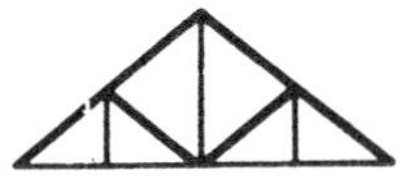

Waddell "A"
(Usually Metal)
Late 19th -
Early 20th century
Length 25-75 feet

Simple Trusses with ancient origin

Town Lattice
(Wood)
Mid-Late 19th century
Length 10-250 feet
Wood diagonals, (no verticals)
in tension and compression

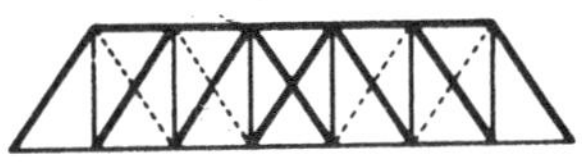

Howe
(Wood, verticals of metal)
19th - 20th century
Length 30-180 feet
Diagonals in compression
Verticals in tension

Burr Arch
(Wood)
1804 - 19th century
Length 50-175 feet
Combination of arch and
multiple King Post

Trusses from the Era of Mostly Wood Construction (End of 19th century)

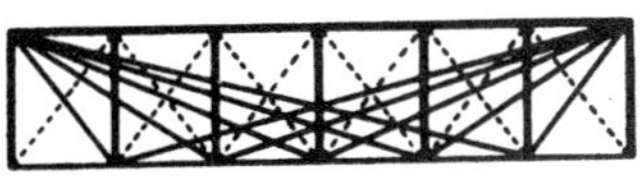

Bollman
1852- Late 19th century
Length 75-100 feet
Verticals in compression
Diagonals in tension
Diagonals from end to each
panel point

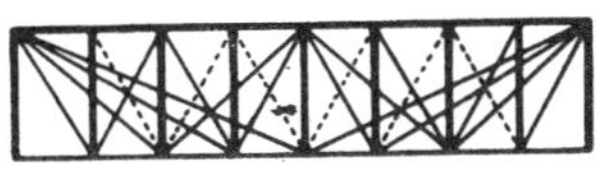

Fink
1881 - Late 19th century
Length 75-100 feet
Verticals in compression
Diagonals in tension
Longest diagonals from end
to center panel point

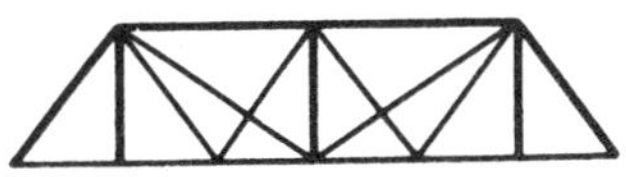

Stearns
1870 - Early 20th century
Length 70-200 feet
Simpler Fink
with verticals omitted at
alternate panels

Trusses based on rod or cable braced diagonals

Figure 2.9a Typical Trusses

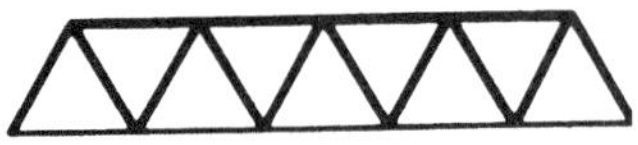

Warren
1848 - 20th century
Length 50-400 feet
True Warren has equilateral triangles in outline
Diagonals carry both tension and compression

Warren with Verticals
Mid 19th - 20th century
Length 50-400 feet
Verticals act as bracing for triangular system
Diagonals carry both tension and compression

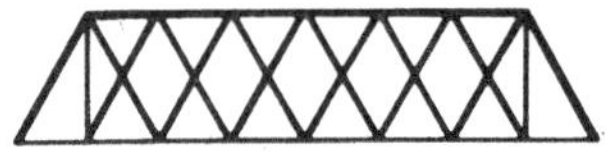

Double Intersection Warren
Mid 19th - 20th century
Length 75-400 feet
Two triangular web systems

Members carry both tension and compression

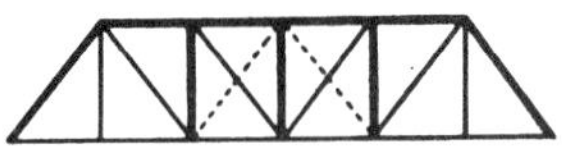

Pratt
1849 - 20th century
Length 20-500 feet
Diagonals in tension, verticals in compression (except near inclined end)

Parker
Mid 19th - 20th century
Length 40-250 feet
A Pratt with Polygonal top chord

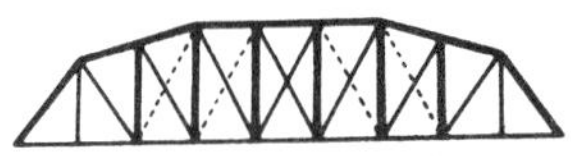

Camelback
Late 19th - 20th century
Length 100-300 feet
A Parker with Polygonal top chord of exactly five slopes

K-Truss
Early 20th century
Length 200-800 feet
Distinctive outline

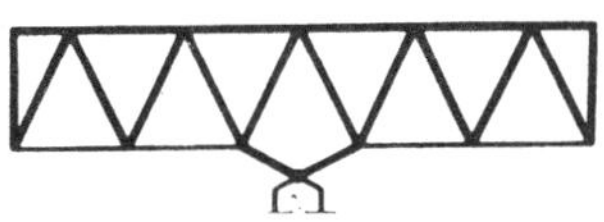

Wichert
1932- Late 20th cnetury
Length 400-1000 feet
Pin Connected support over piers

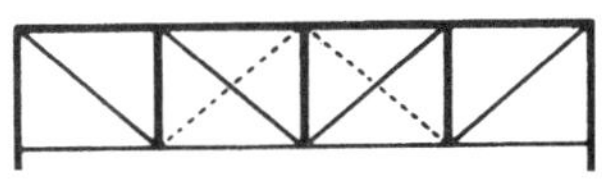

Truss Leg Bedstead
Late 19th - Early 20th century
Length 30-100 feet
Pratt with vertical end posts imbedded

Trusses Constructed of Metal

Figure 2.9b - Typical Trusses - Taken from poster by Historic American Engineering Record Library of Congress

Chapter 3 - Basic Principles of Trusses

Triangles are used extensively in truss design. In Figure 2.9, note the patterns of triangles in the truss outlines. In the King Post and Warren Trusses the triangles are obvious. In other designs such as the Fink and Bollman Trusses, the diagonals form triangles which overlay a system of verticals. The use of triangles in the analysis of trusses is covered in more detail in Chapter 6.

3.1 The Triangle Building Block

The triangle is the fundamental building block in structures. It is the strongest formation of members with joints. Pinned structures can collapse if a load is placed on a joint which is not supported by a member. With triangles a load applied in any direction always has a member at some angle helping brace against the load. With other shapes, a pinned structure may be able to rotate at a pin with no resisting member. Sketches of triangles and other shapes shown below help demonstrate this principle (in this example, the bottom member is attached to something solid like the floor but each joint is a pin which can rotate freely, like a single bolt through boards) A typical deflection is shown in the same diagram. With the square and 5-sided polygon, no member can brace the load and large deflections are possible.

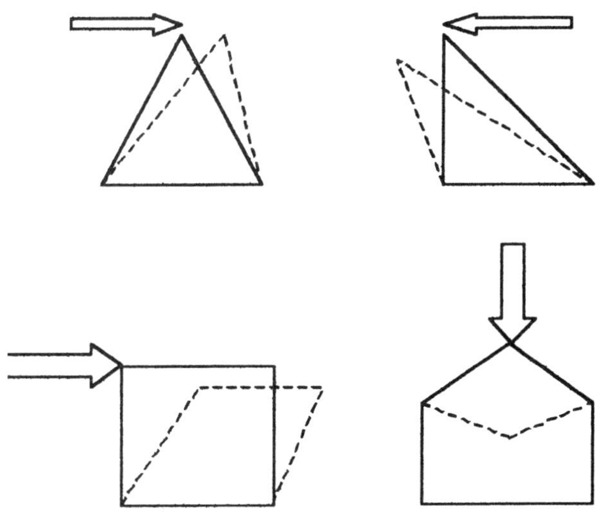

Figure 3.1 - Shapes with Loads

3.2 Loads in Truss Members

Loads in trusses, known as structural forces, are carried along the members of the truss. Loads of this type are referred to as normal loads. A normal load will produce either a tension - a force which lengthens the member or tries to pull the member apart, or a compression force - a force which shortens the member.

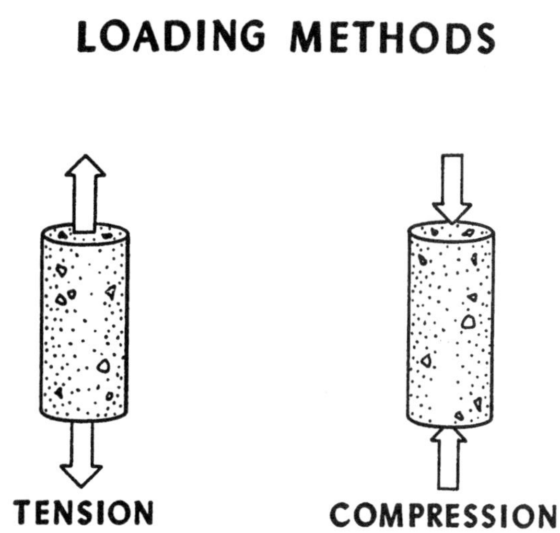

Figure 3.2 -Loads - Tension and Compression

Many of the early trusses are considered as simple trusses. Simple trusses are assumed to carry only normal forces. No twisting (moments) or other forces are considered in their design. The use of a pin where members were joined meant that any load which tried to twist a member would simply rotate the member around the pin.

Gusset plates use bolts, rivets, and welded connections at the ends of members which fixes ends and allows forces which twist, moments, to be carried. In general, these are considered negligible for the purpose of the analysis of trusses. As a matter of safe design, this assumption needs to be checked carefully. A structure which can resist moments is referred to as a frame. Frame analysis may need to be completed as appropriate for the loading within the truss members.

Figure 3.3 A bridge under load by a passing train

3.2 External loads on trusses

Loads are caused on the structures that we build in many ways. Objects, of course, have weight due to gravity, which means the members of a structure create a load spread out over all members. Loads caused by the structure which are always present are known as dead loads.

Loads which come and go during the life of a structure are known as live loads. Good examples of live loads are trains or other traffic passing over a truss.

Various other kinds of loadings such as snow, roadbeds and other loads create a load over all or some of the structure. A load which acts over all, or a good part of the structure is called a Distributed Load.

Concentrated Loads are caused by loads that act on a very small area or on a point in the structure. Dead loads and live loads may be concentrated or distributed loads based on their source.

When designing structures, engineers receive guidance in the form of standards for loads which are common on the structures. Unique loads must also be considered for each structure designed. This book is not intended for use as a design guide for structures, but example loads are shown in the following tables and figures.

3.2.1 Typical Railroad Loadings

Perhaps the most famous and common loading for railroad bridges was devised by Theodore Cooper. The Cooper Load, commonly referred to as the E loadings, represent two locomotives followed by a line of freight cars. A series of concentrated loads is used for the locomotives, while a distributed load is used to represent freight cars. Figure 3.4 shows two typical locomotive loads from designs of 1888. The wheel loading of the train is used in the analysis of trusses to find the forces in the members. The generalized Cooper loading would come into use in 1894 and is shown in Table 3.1. As an example of the format in Table 3.1, the 88 ton Consolidation Engine loadings are shown in the same format. Note the change in the table from tons to the load unit kips. Since loads can be large in structures, a notation of 1 kip being equal to 1,000 pounds is often used. Typical of the time was an axle loading of 40 kips on the drive axle, this loading system was subsequently referred to as the E-40 loading. Since the introduction of this system of loading, trains have increased in size. The E loading for more contemporary trains may be obtained by proportioning the locomotive axle loads for the larger engines. In today's design, loads of E-72 or E-90 are common.

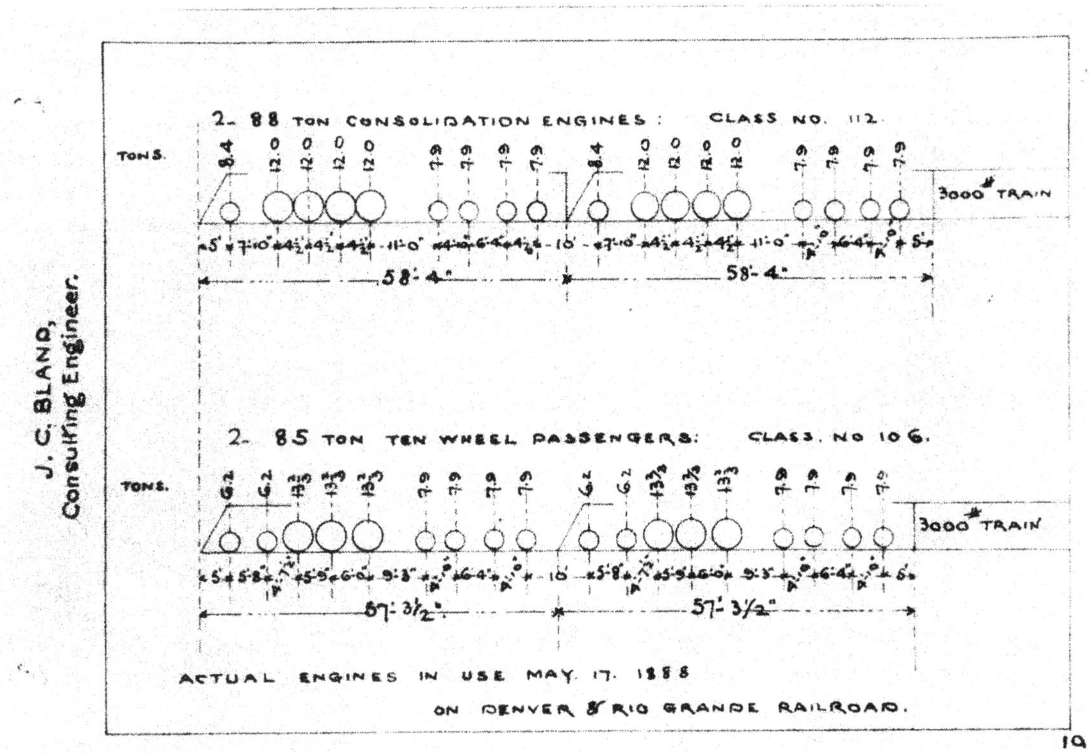

Figure 3.4 - 1888 Railroad Design Loads
Courtesy, Colorado Historical Society

Loading Tables:

88 ton above(example)

	1st Engine / 2nd Engine / Freight Cars
Spacing to next axle (Ft.)	7 4 4 4 11 4 6 4 10 7 4 4 4 11 4 6 4 5
(In.)	10 6 6 6 10 4 10 10 6 6 6 10 4 10
Load in tons	8.4 12 12 12 12 7.9 7.9 7.9 7.9 8.4 12 12 12 12 7.9 7.9 7.9 7.9 1.5 tons/ft.

Cooper (E) standard design loading

Spacing to next axle (ft.)	8	5	5	5	9	5	6	5	8	8	5	5	5	9	5	6	5	5	5	
E40 (kips)	20	40	40	40	40	26	26	26	26	20	40	40	40	40	26	26	26	26	26	4 kips/ft.
E72 (kips)	36	72	72	72	72	47	47	47	47	36	72	72	72	72	47	47	47	47	47	7.2 kips/ft.
E80 (kips)	40	80	80	80	80	52	52	52	52	40	80	80	80	80	52	52	52	52	52	8 kips/ft.
E90 (kips)	45	90	90	90	90	59	59	59	59	45	90	90	90	90	59	59	59	59	59	9 kips/ft.

Table 3.1 Cooper loadings for railroad bridge design

3.2.2 Typical Highway Loadings

Standard highway loadings have also been devised for the design of highways (and may change over time). Loads for highway bridges are based primarily on the largest loads which may be generated by truck traffic. For smaller bridges a two axle truck may be used which can have either 20, 15, or 10 kips as the total weight of the truck. These standard loadings are known as the H20, H15, and H10 trucks. The front axle carries two-tenths of the total weight, while the rear axle, placed 14 feet on center from the front axle, carries eight-tenths of the weight. Examples of these standard truck design loadings are shown in Figure 3.5 Any assumptions made about the type of truck which may be carried on a particular bridge will be posted as a weight limit at the bridge entrance by the appropriate highway department responsible for the bridge.

For longer bridges, a two axle truck with one semi-trailer axle is used, which represents larger semi trailer trucks. One such design truck is the H20-S16 truck. Like the previous standard, the truck tractor weight is proportioned between two axles, 4 kips to the front axle and 16 kips to the rear axle. An additional 16 kip load is placed at varying distance of 14 to 30 feet behind the tractor loads (whichever distance generates the highest load on the truss). A distributed load, a uniform lane load which represents a line of cars halted on the bridge is also considered, as is specified in standards. The maximum force created in any member will be the basis for the design of that member.

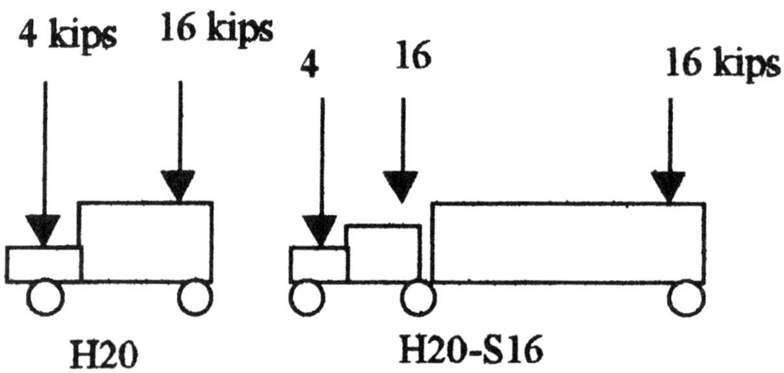

Figure 3.5 Typical Highway Loads

3.2.3 Typical building loads

Trusses are also used in buildings to support floors or roofs. For building designers, design guides are available for both dead (construction) loads and in-use (live) loads. Dead loads are determined primarily by the material used. Live loads are determined by the use of the structure. Typical values are shown in Table 3.2.

Table 3.2 Typical Building Loads

Dead Loads

	Pounds per Square Foot
Concrete	144
Floor Construction	
Joists	6
Oak Floor	45
Brick	130

Live Loads

	Pounds per Square Foot
Armories	150
Foundries	600
Factory	125
Warehouse	125
Office	80
Homes	40

Chapter 4 -Some Real Bridges

4.1 Bollman Pinned Truss

Figure 4.1 Photo of Bollman Truss Bridge
Photo reproduced from Historic American Buildings Survey/ Historic American Record, Library of Congress

This account is taken from text available on-line from the Library of Congress, Historic American Buildings Survey/ Historic American Engineering Record. Additional text, and references may be found on this web site.

The Bollman Truss bridge is credited as being the first type of iron truss bridge used in large numbers on American railroads. The bridge shown above, erected in 1887, spans the Little Patuxent River at Savage, Howard County, Maryland. It was the last erected of this type. The bridge was designed by its namesake, Wendel Allen Bollman, when he was working for the Baltimore and Ohio Railroad.

The bridge is located at the end of a now abandoned B&O spur constructed to serve the cotton duck mill at Savage. It consists of two spans. The design uses diagonal tension members of wrought iron and vertical compression members of cast iron. The floor beams of each panel are supported independently by two diagonal tension members that are connected to the vertical end posts.

Howard County has restored the bridge for use in a recreational site. This will preserve an example of one of America's oldest standing railroad bridges.

Selected drawings for use in constructing models of this or similar structures are contained on the following pages. The drawings are produced directly from the originals to maintain historical accuracy.

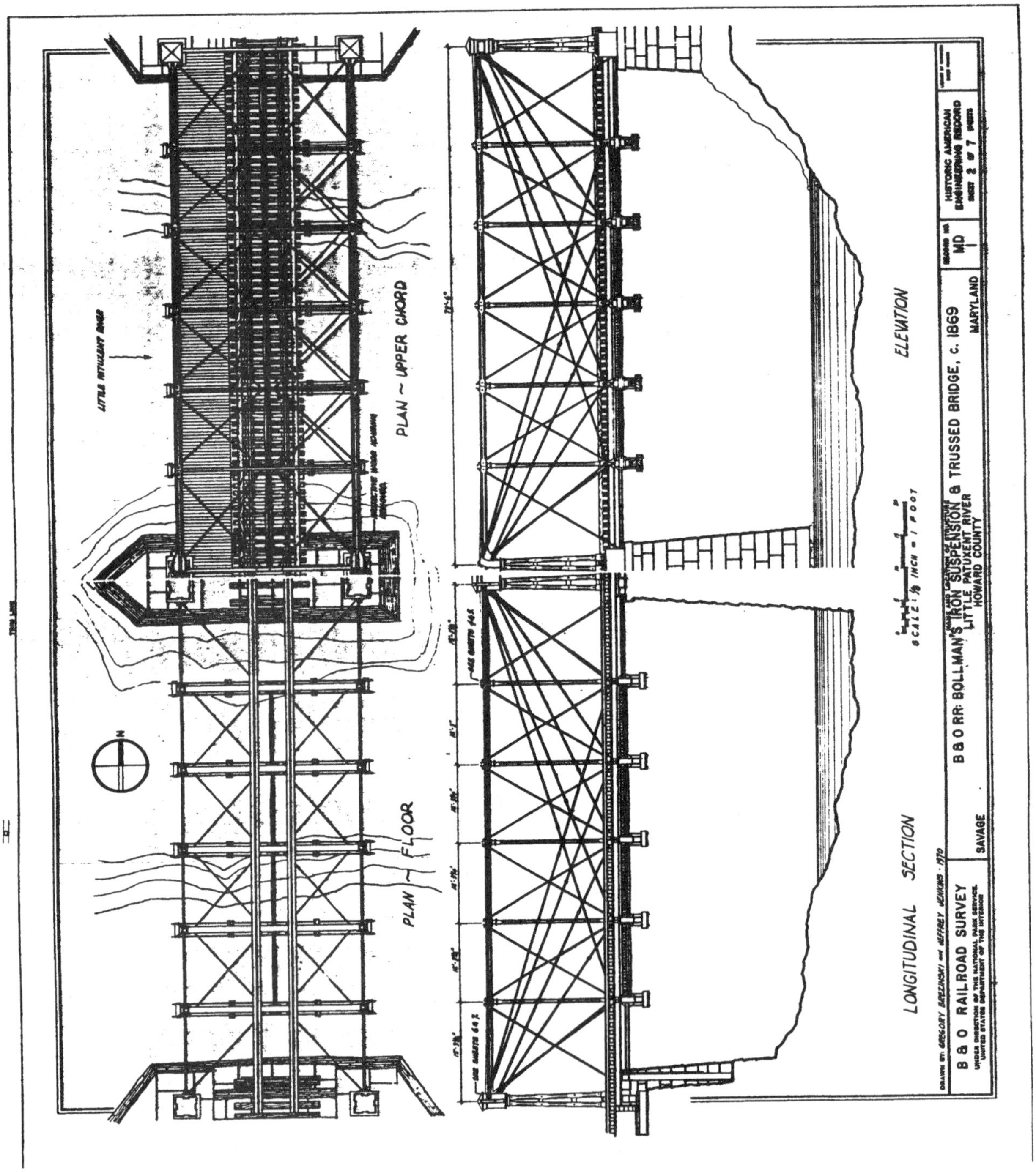

Figure 4.2 - Bollman Truss Bridge - Longitudinal Section
Reprinted from original of the Historic American Buildings Survey/ Historic American Record, Library of Congress

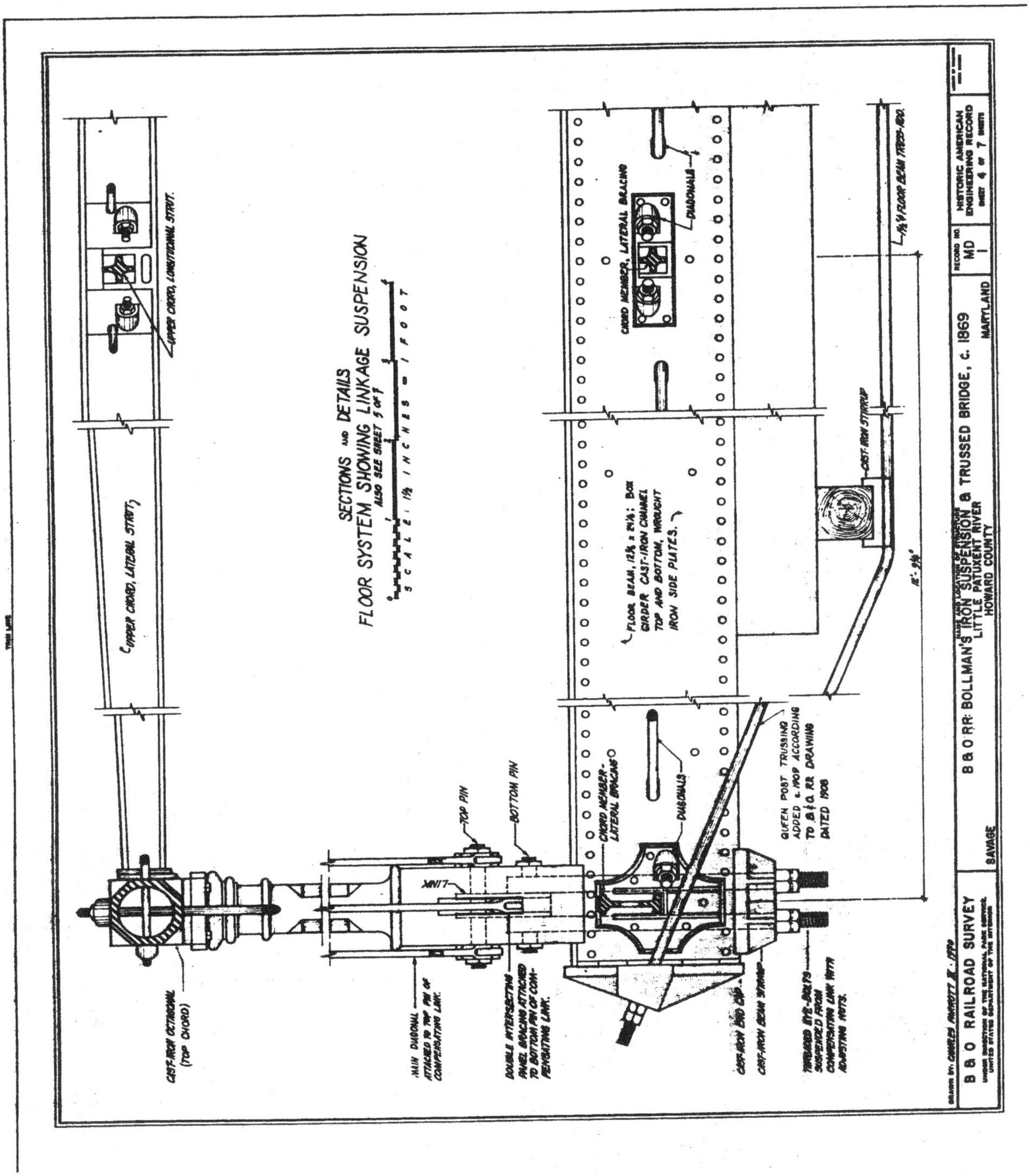

Figure 4.3 - Bollman Truss Bridge - Floor System Linkage
Reprinted from original of the Historic American Buildings Survey/ Historic American Record, Library of Congress

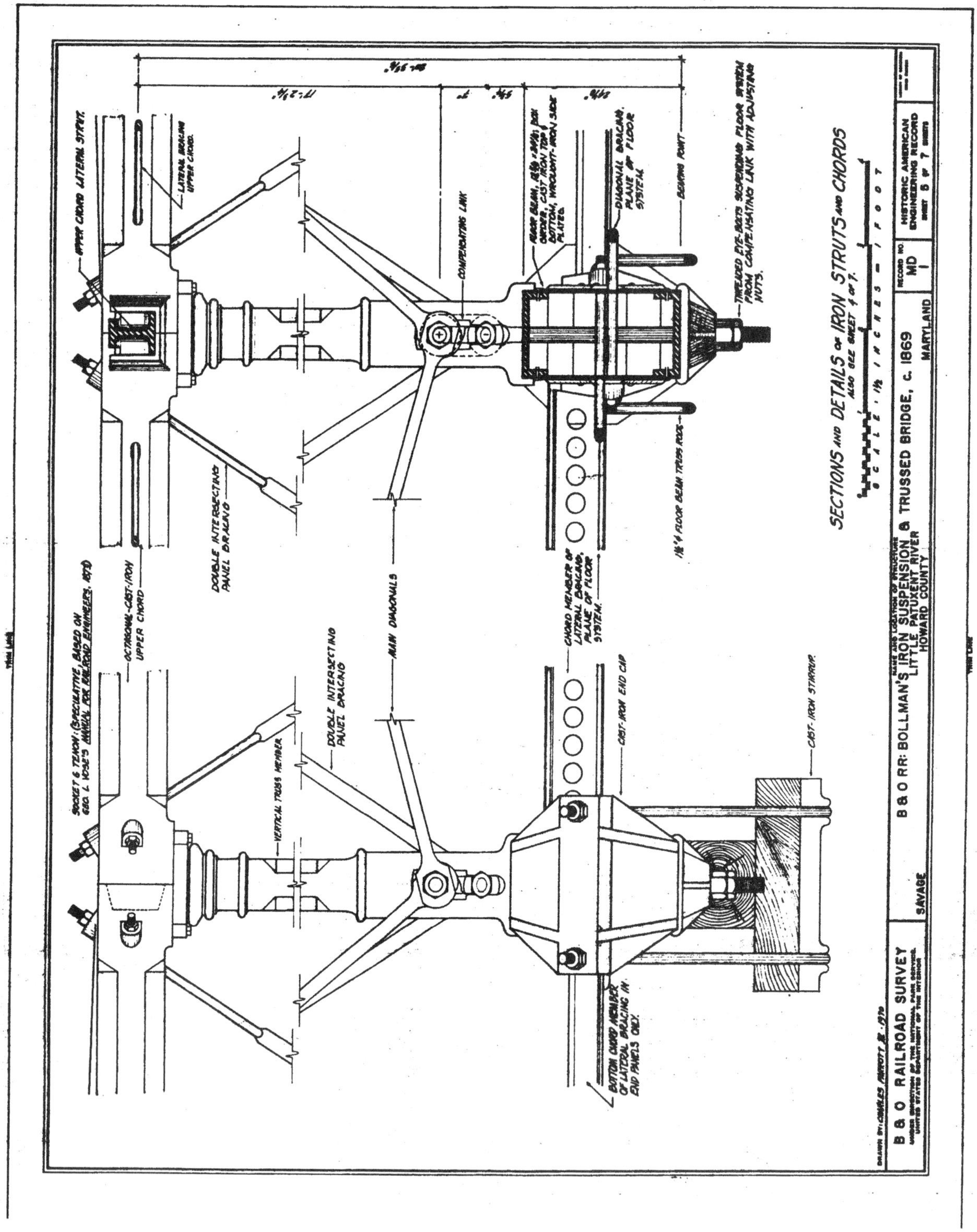

Figure 4.4 - Bollman Truss Bridge - Details of Iron Struts and Chords
Reprinted from original of the Historic American Buildings Survey/ Historic American Record, Library of Congress

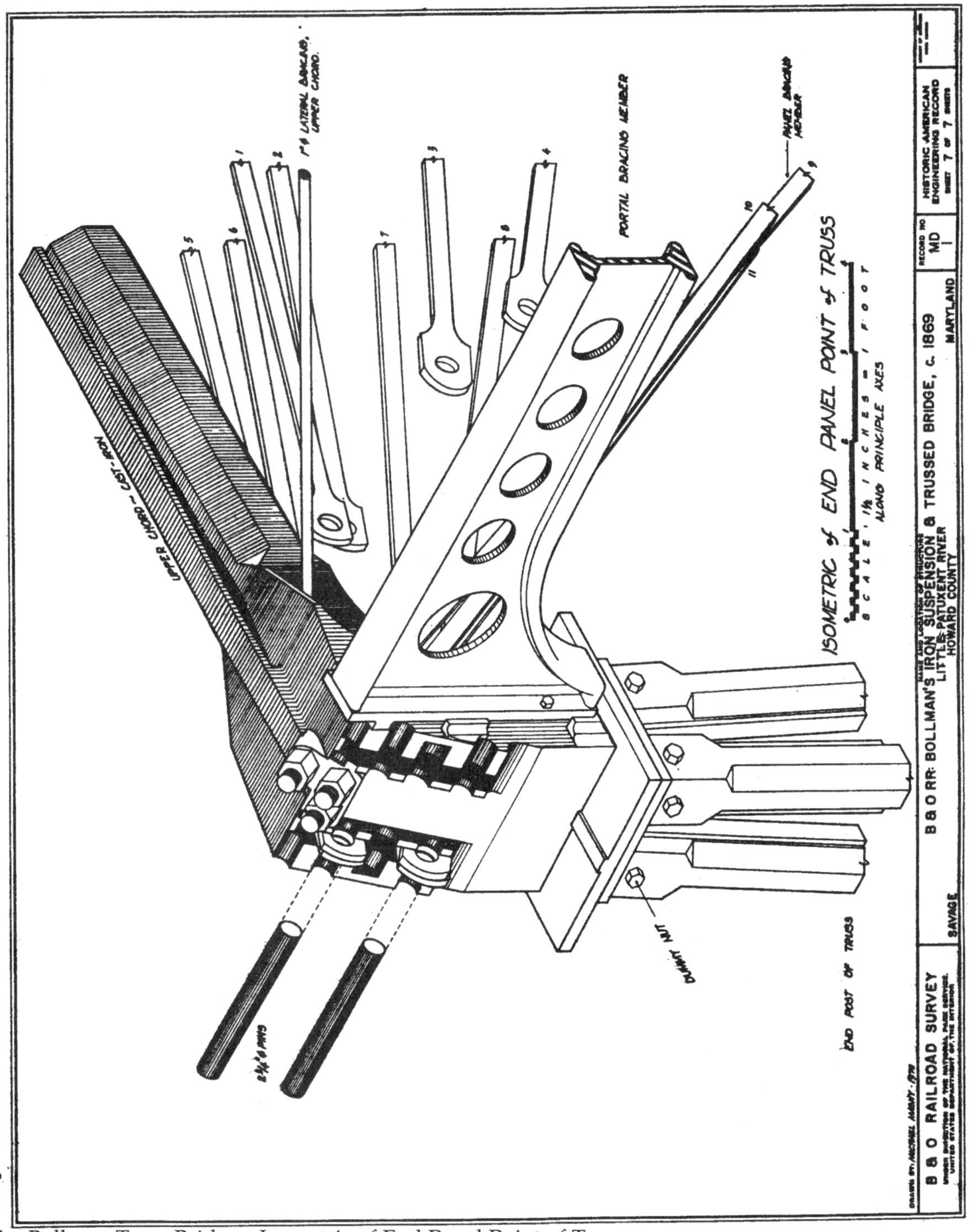

Figure 4.5 - Bollman Truss Bridge - Isometric of End Panel Point of Truss
Reprinted from original of the Historic American Buildings Survey/ Historic American Record, Library of Congress

4.2 Warren Truss

Figure 4.6 Photograph of the bridge across Big Creek, Tennessee
Reprinted with permission of the Association of American Railroads

Located near the town of Del Rio in the southeast corner of Tennessee is an open deck Warren Truss railroad bridge. This bridge, built in 1919, is a single track bridge on the Norfolk Southern line at milepost 195.3. The line runs from Knoxville, TN to Asheville, North Carolina. The bridge carries mostly coal train traffic with mixed freight making up the majority of the remaining traffic

The truss spans approximately 160 feet across Big Creek. The chord members and diagonals are latticed built up sections. The vertical and floor members are built up sections as shown in the following figures. The floor system to support the rails, see Figure 4.7, is set on 6ft-6in centers and spans 26 feet ½ inch between the trusses.

Of particular interest are the excellent line diagrams and cross sections prepared for the study. These simple diagrams, which follow, provide a good understanding of the construction of bridges. Figure 4.8 shows the plan and elevation views of the bridge. Figures 4.9 through Figure 4.12 show details of the construction of all members within the bridge.

In 1993, the Association of American Railroads Technical Center in Chicago, Illinois did an extensive study of the loadings and effects on this bridge. In general, loads were in the expected 60 (Cooper E60) to 70 kips range. Stresses(see Chapter 7)were in the 3500 psi range.

Figure 4.7 Floor system of Warren truss

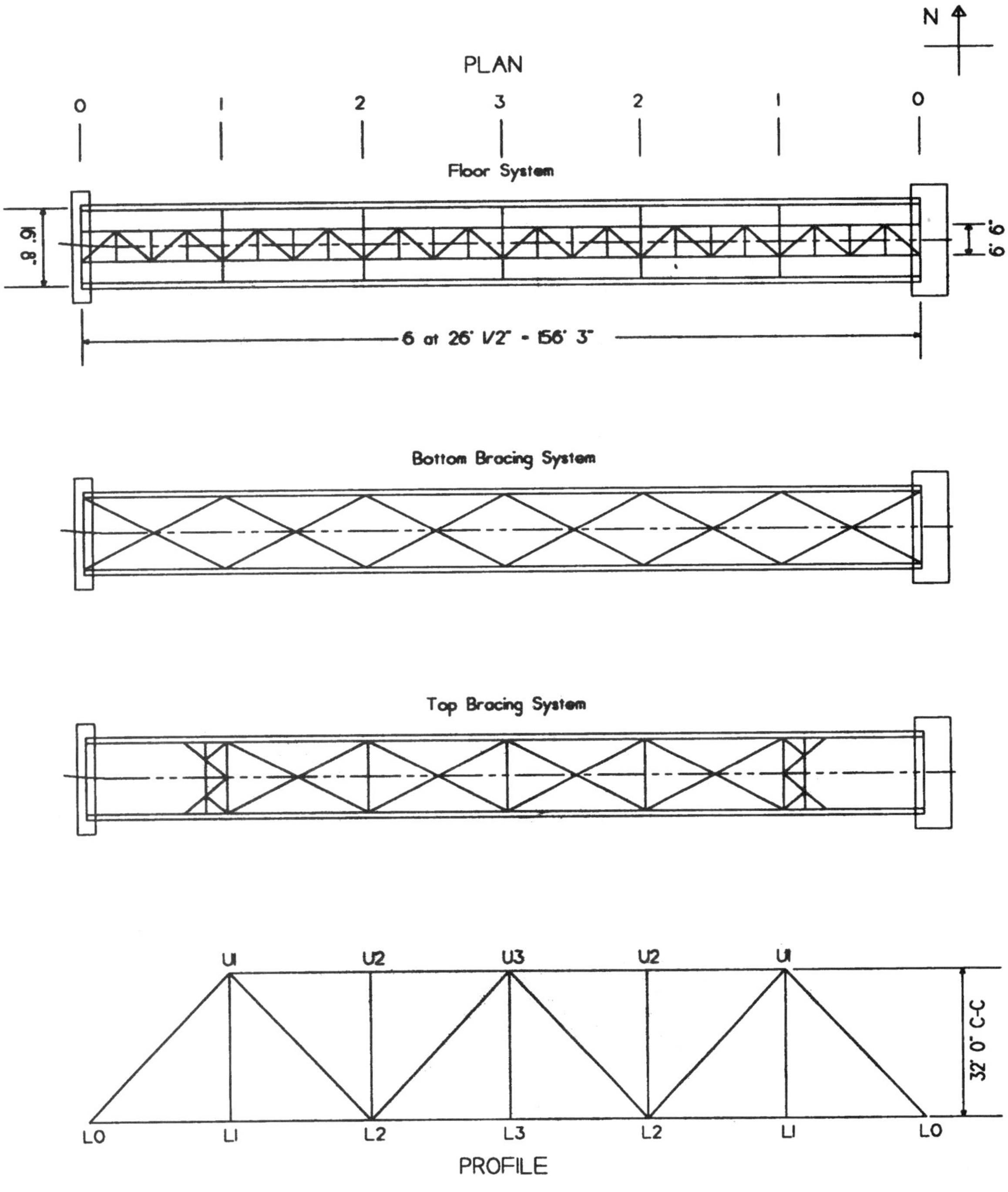

Figure 4.8 Plan and elevation of a Warren Truss Bridge
Reprinted with Permission of the Association of American Railroads

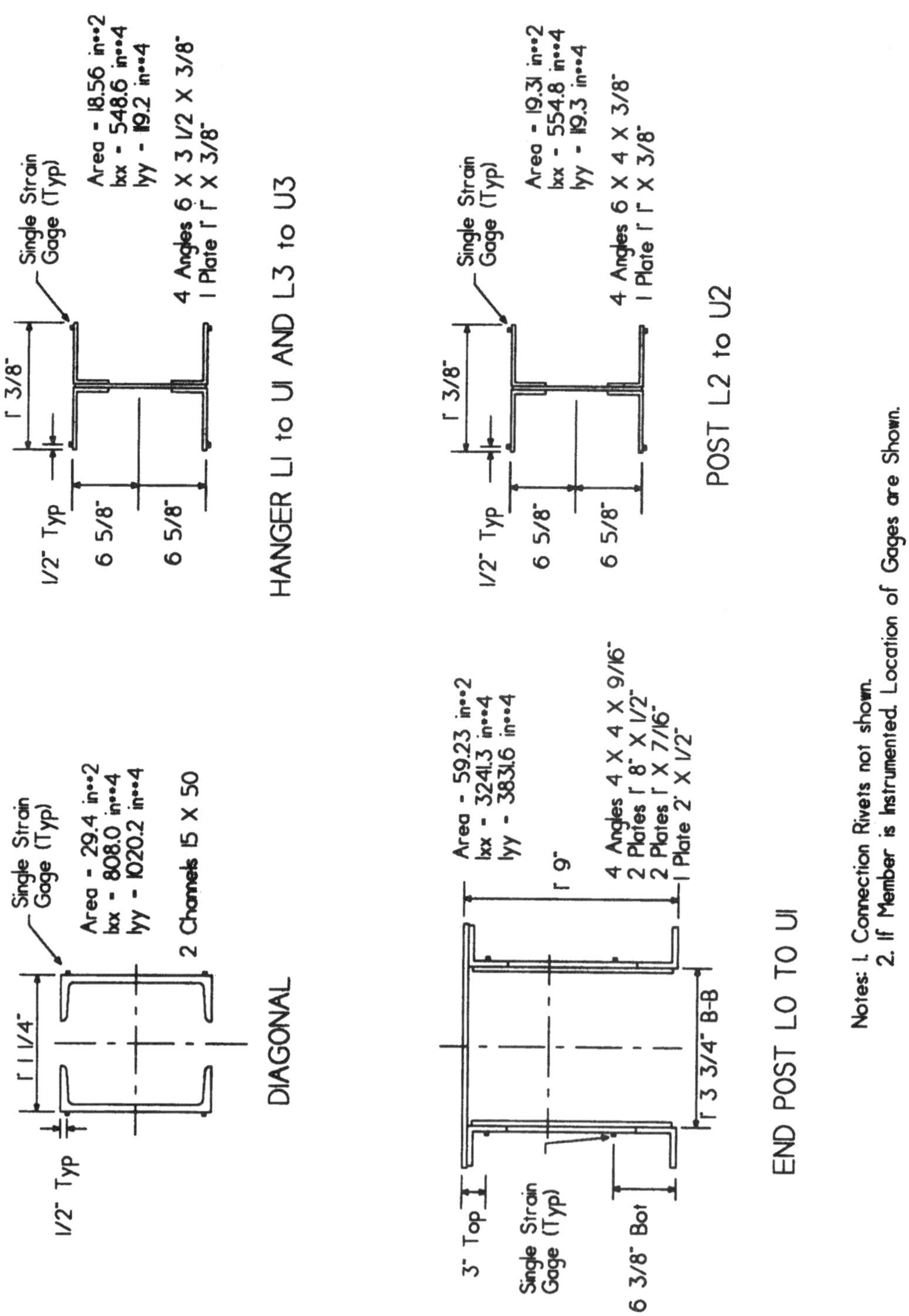

Figure 4.9 Cross sections of bridge members
Reprinted with the permission of the Association of American Railroads

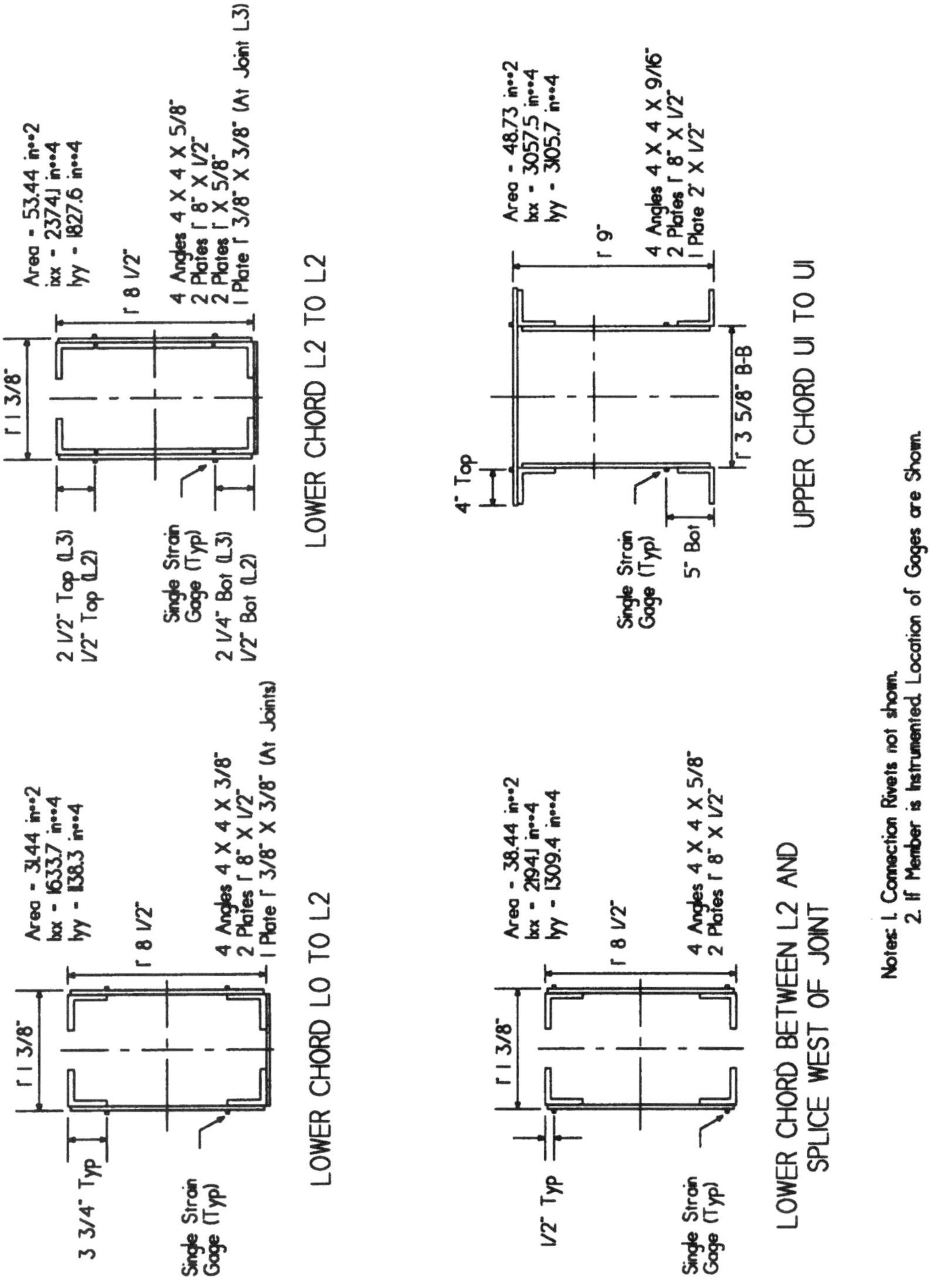

Figure 4.10 Cross sections of bridge members
Reprinted with permission of the Association of American Railroads

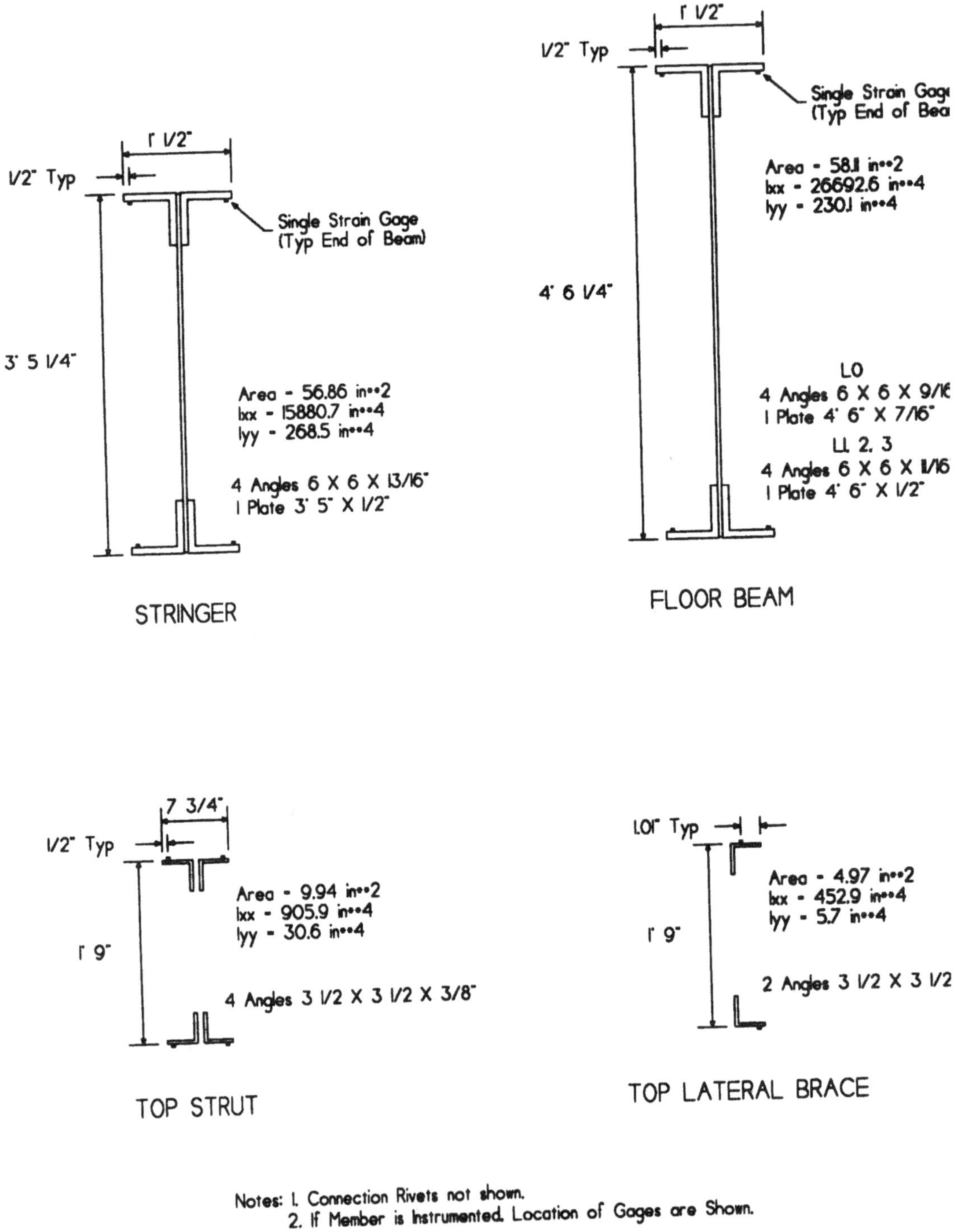

Figure 4.11 Cross Sections of bridge members
Reprinted with permission of the Association of American Railroads

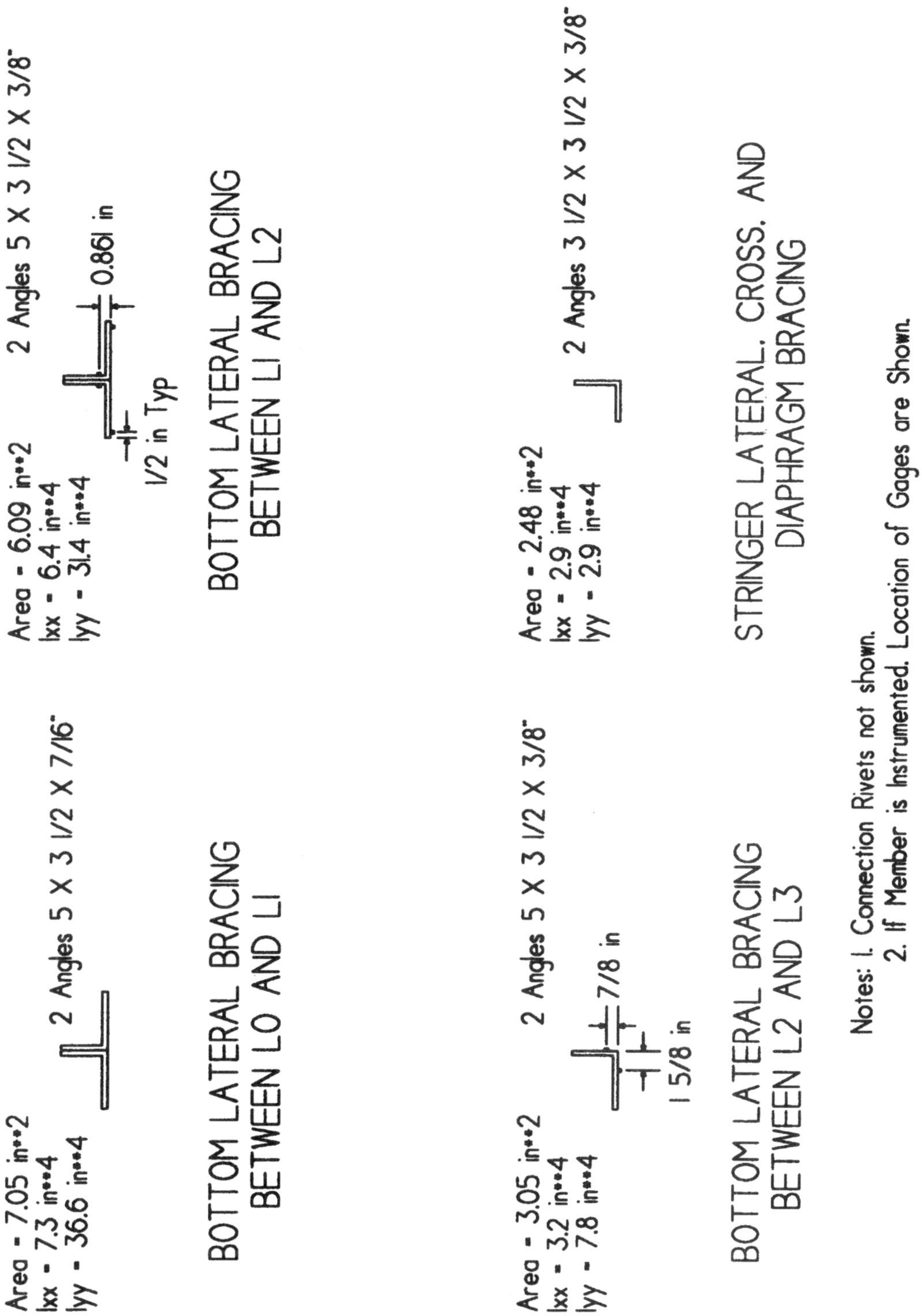

Figure 4.12 Cross sections of bridge members
Reprinted with permission of the Association of American Railroads

4.3 Howe Truss

There is always a trade-off in engineering design between the cost efficiency of using standard designs repetitively and the need for custom and site specific designs. As early as 1888, the Denver and Rio Grande Railroad was beginning to use standard designs for bridges and making site specific adjustments in the approaches to the bridge and the river piers. Figure 4.13 shows a short span Howe truss in use. Figure 4.14 shows a standard drawing for the Denver and Rio Grande Railroad Standard 48 foot Howe Truss. The track placement is shown in Figure 4.15. In this design, special devices were prefabricated for the joints, these were called seat blocks. A typical seat block detail, is shown in Figure 4.16. The drawings are reproduced directly from original blueprints for historical accuracy. The dates from the photo in Figure 4.13 indicate that the bridge is not constructed from the following blueprints, it is just an example

Figure 4.13 Bridge near Pueblo, CO late 1890's
Courtesy: Colorado Historical Society

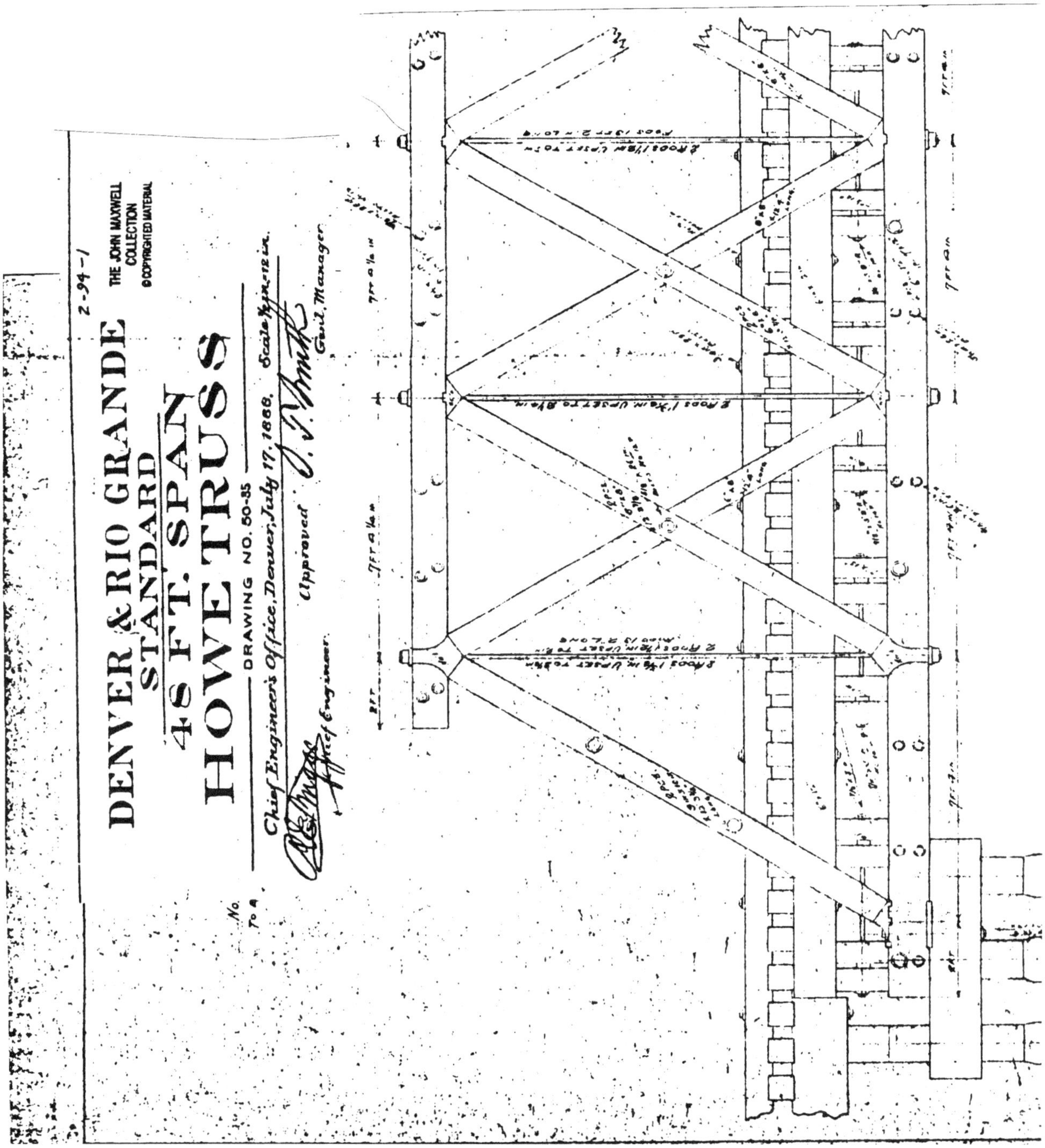

Drawing notes are shown in Chapter 5

Figure 4.14 - Denver Rio Grande Standard 48 foot Howe truss
Reprinted with Permission: John Maxwell Collection, 1055 Jessica, Fernley, NV

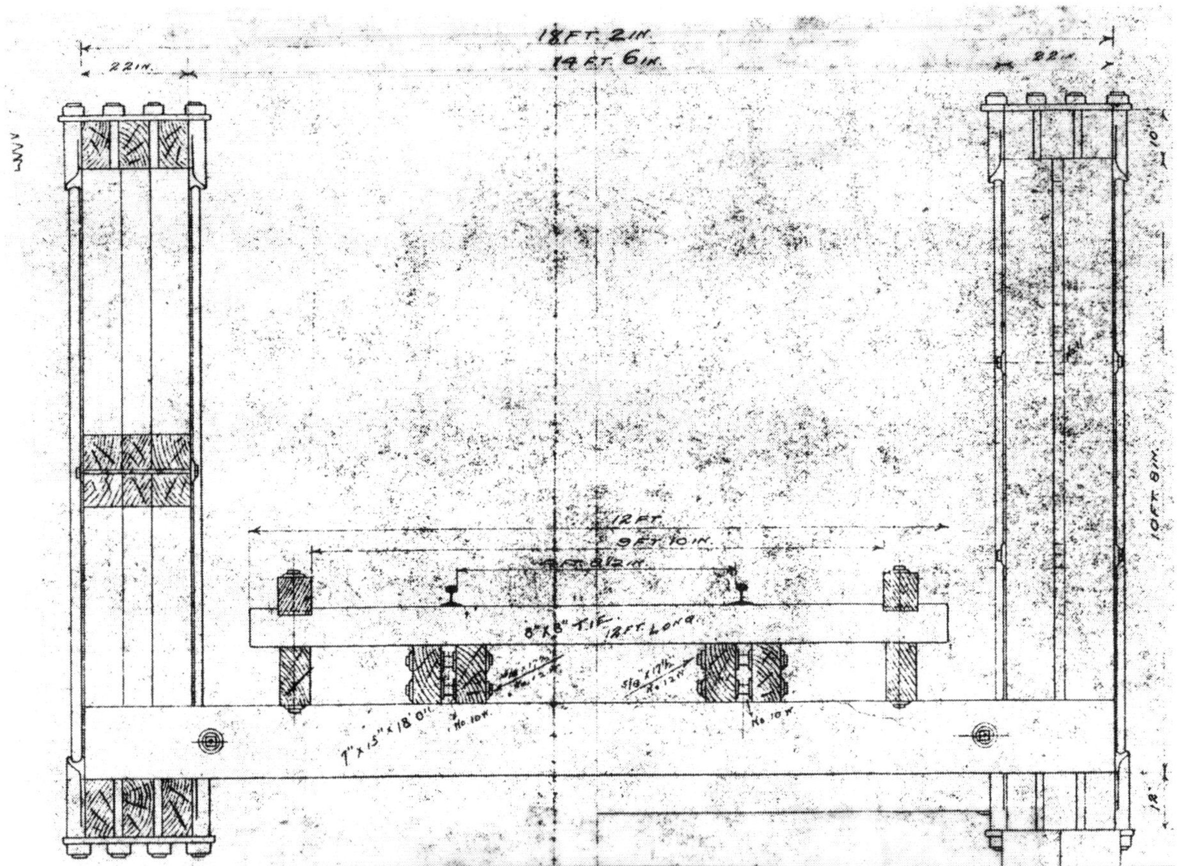

Figure 4.15 - Cross Section of Standard Howe Truss
Reprinted with Permission: John Maxwell Collection, 1055 Jessica, Fernley, NV 89408

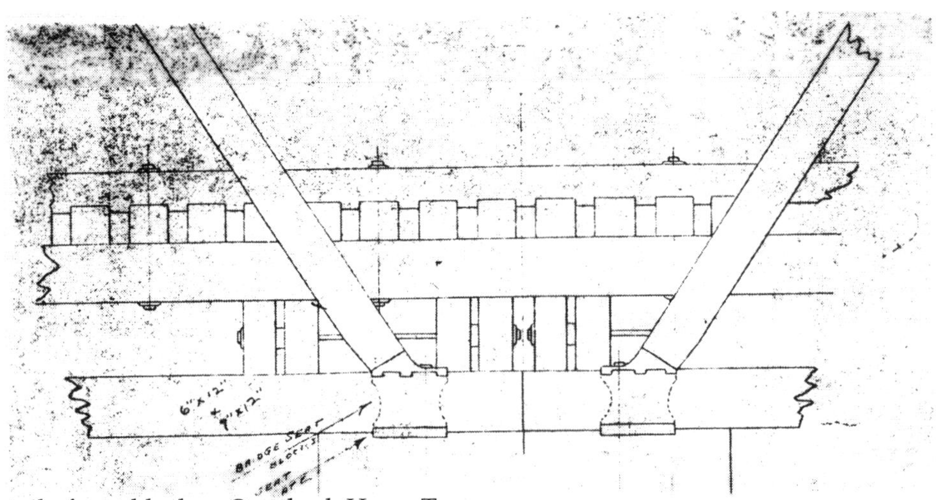

Figure 4.16 - Detail of seat blocks - Standard Howe Truss
Reprinted with Permission: John Maxwell Collection, 1055 Jessica, Fernley, NV 89408

4.4 Pratt Truss

Figure 4.17 - Cherry Creek Railroad Bridge
Photo from the On-Line Library of Congress Collection HABS/HAER Collection

This account is taken from text available on-line from the Library of Congress, Historic American Buildings Survey/ Historic American Engineering Record. Additional text, and references may be found on this web site.

The single span pinned through Pratt Truss was a commonly used bridge in the late 19th century. An example of this bridge type is shown in Figure 4.17. Although now abandoned, this bridge has been saved for historical reference and is proposed for possible reuse by the City of Denver. The current owner is the Burlington Northern Railroad.

Denver's railroad era began in 1870 with the arrival of the Denver Pacific and Kansas Pacific Railroads. These lines provided a connection to the transcontinental line which had passed north of Denver going through Cheyenne. During the 1880's railways prospered along the S. Platte River valley and provided industrial delivery to warehouses located in lower downtown and West Denver.

The Cherry Creek bridge was constructed by the Colorado and Southern company. This 105' span bridge was probably built in 1890. The bridge consists of five panels, each 20 feet, 10 inches long and is comprised of Carnegie steel or wrought iron components. Pinned connections are used to connect the major structural components. Channel sections with riveted lacing bars and cover plates are utilized for the upper chords, end posts and verticals. This construction is similar to examples shown in Figures 4.9 and 4.10. The lower chords and outer panel diagonals consist of flat steel with holes provided on each end to form the pinned joint. The center diagonals and top and bottom lateral bracing members are similar but are made from square stock and contain turnbuckles, a device that can be turned to tighten the member.

The bridge does have some unusual decorative portal bracing. At either end of the bridge a metal plate is riveted between the end posts and portal strut with quatrefoil pattern, see top member Figure 4.18. Such decorative trim was characteristic of the late 19th century bridges.

Figure 4.18 East Portal of Cherry Creek Bridge
Photo from the On-Line Library of Congress Collection HABS/HAER Collection
Photographer: Clayton B. Fraser

4.5 Waddell A Truss

Figure 4.19 Photo of Waddell A Truss Bridge
Photo reproduced from Historic American Buildings Survey/ Historic American Record, Library of Congress

This account is taken from text available on-line from the Library of Congress, Historic American Buildings Survey/ Historic American Engineering Record. Additional text, and references may be found on this web site.

In 1898, the Quincy, Omaha and Kansas Railway built the Linn Branch Creek Bridge in accordance with the original patent filed by John Alexander Low Waddell. This bridge type was constructed for relatively short spans and used pins instead of rivets in construction. This construction method was widely used in the Midwest and in Japan. In 1953 this bridge was converted for use as a highway bridge and was used for approximately 25 years until its removal.

J.A.L. Waddell was among the most respected American civil engineers of the late 19th and early 20th centuries. During his life, he was involved in the design and construction of hundreds of bridges. During his career Waddell was critical of the use of trusses that were not braced on the top chord. These are known as Pony Trusses. In the 1880's a substantial number of rivets were required in the field in less than optimal conditions. The construction of joints could be questionable if not riveted properly (see Chapter 8). The Waddell "A" truss solved both of these problems by being sufficiently tall to allow lateral cross bracing, and being pinned. As the riveted connection emerged, Waddell encouraged the use of riveting machines operated by trained workers. Note that the form follows the King Post truss from the middle ages.

A complete set of drawings for this bridge appears on the next three pages.

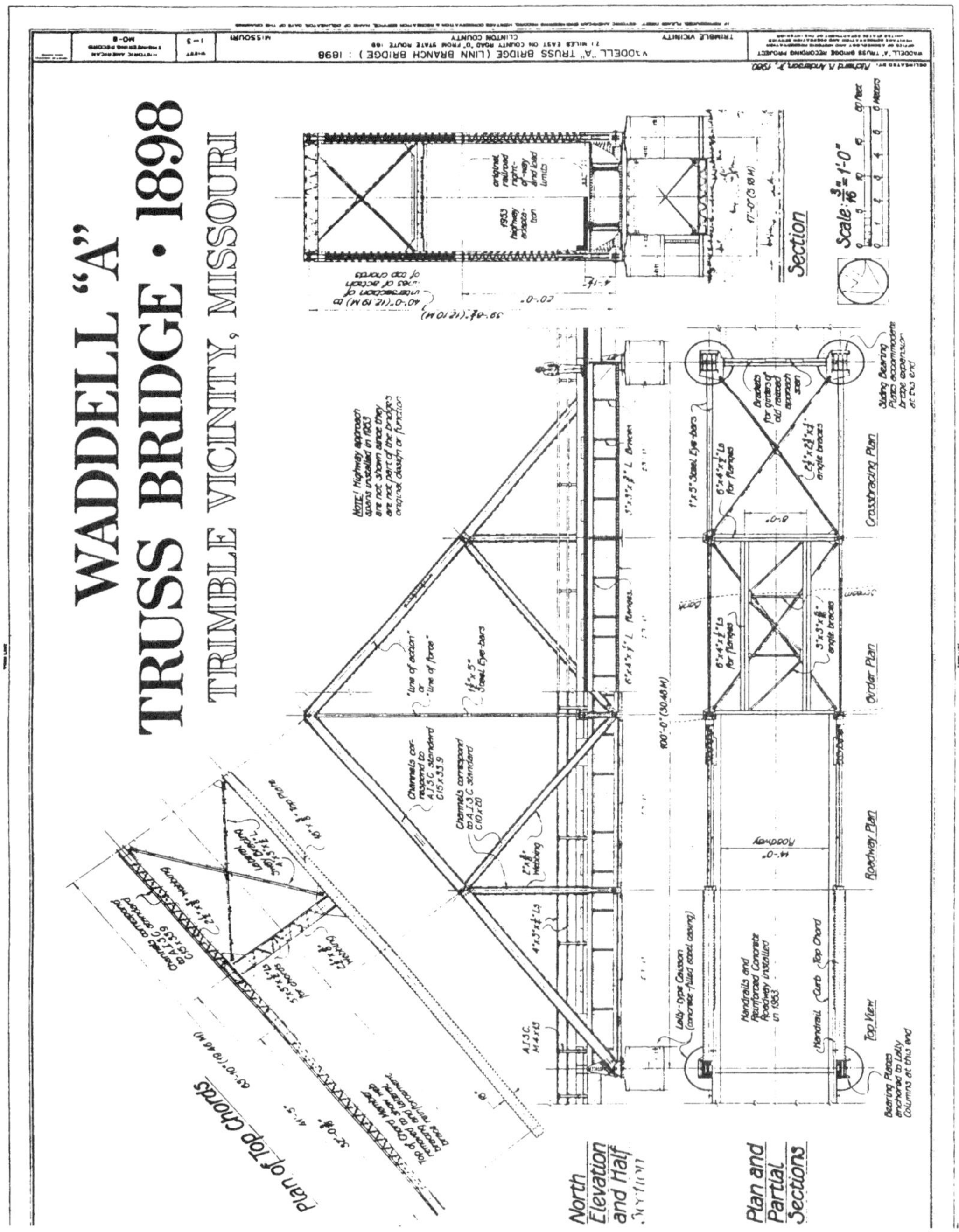

Figure 4.20 - Waddell A Truss Bridge - Plan and Partial Sections
Reprinted from original of the Historic American Buildings Survey/ Historic American Record, Library of Congress

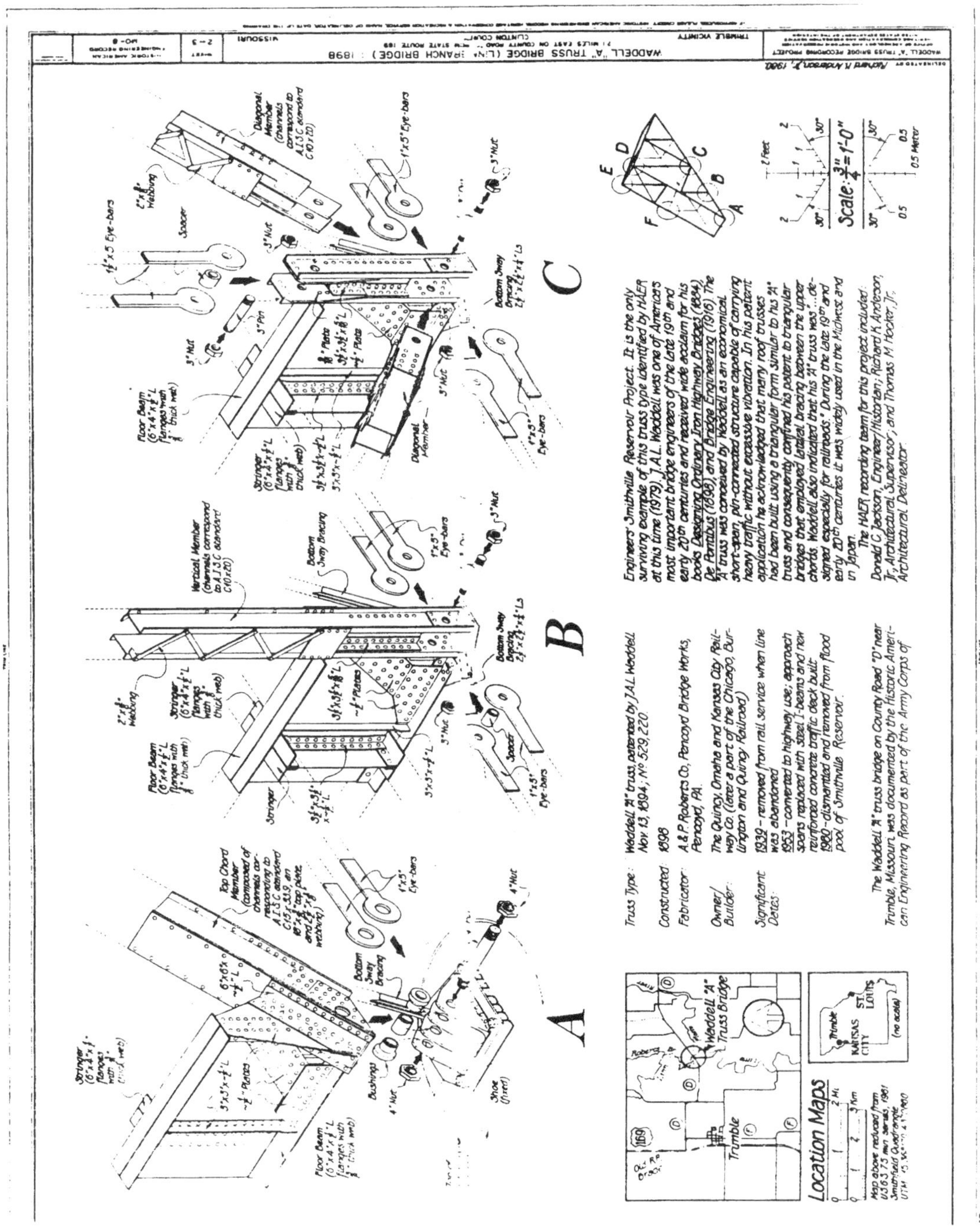

Figure 4.21 - Waddell A Truss Bridge - Sections A-C
Reprinted from original of the Historic American Buildings Survey/ Historic American Record, Library of Congress

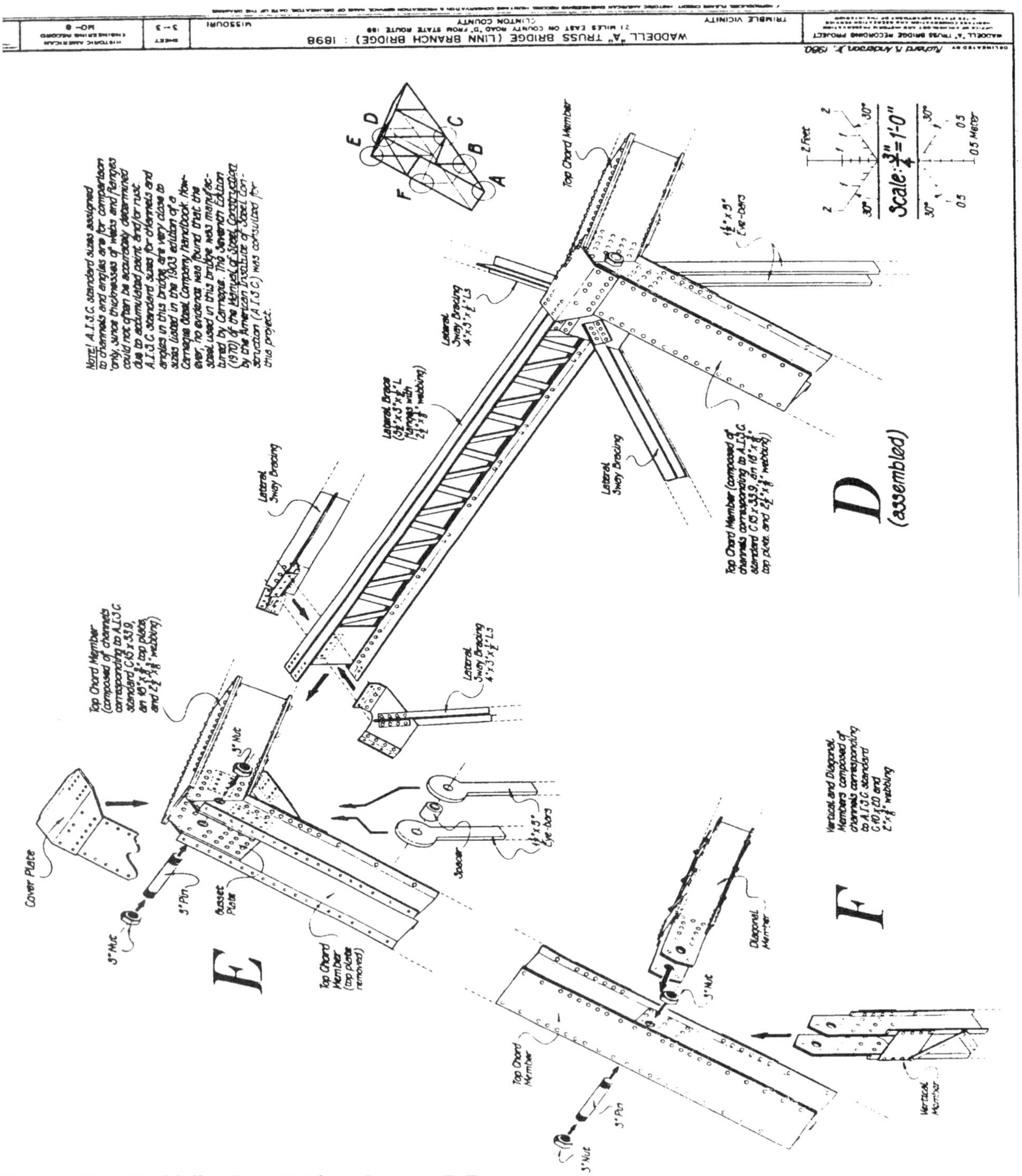

Figure 4.22 - Waddell A Truss Bridge - Sections D-F
Reprinted from original of the Historic American Buildings Survey/ Historic American Record, Library of Congress

4.6 Lyons Brook Truss and Plate Girder Bridge

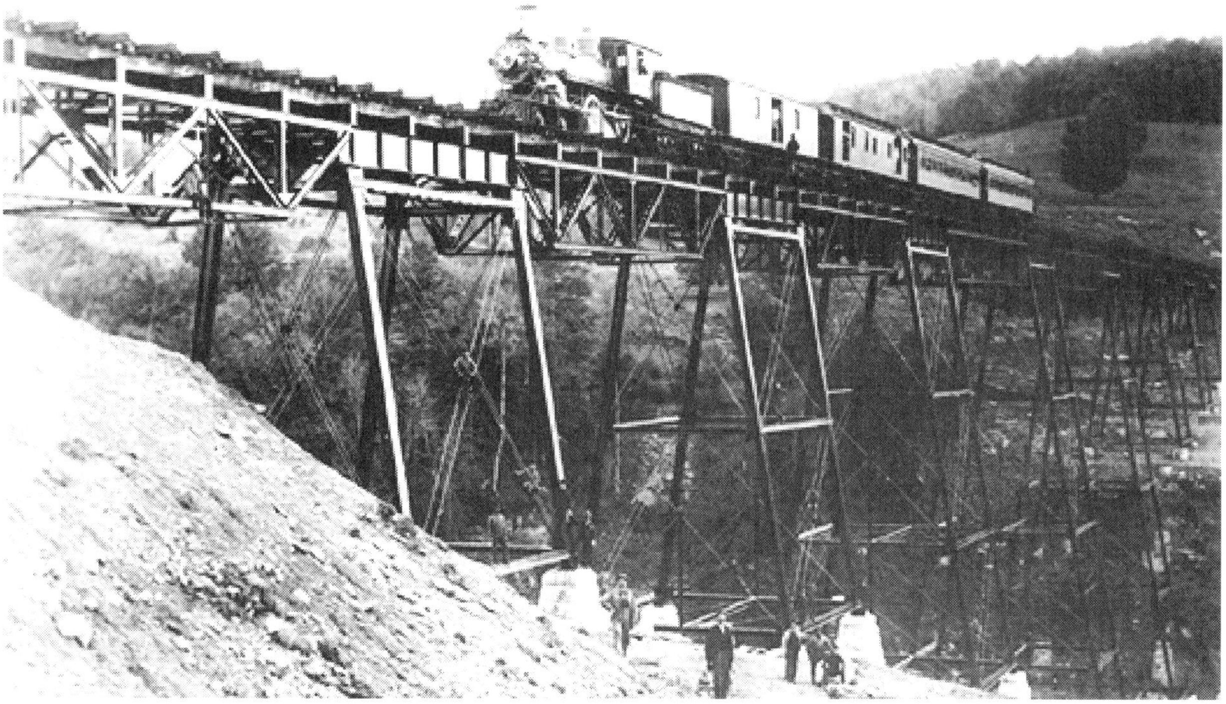

Figure 4.23 - The Lyons Brook Bridge
Used with permission of the Ontario & Western Historical Railway Society

(Text excerpted from the Ontario& Western Historical Railway Society Members On-Line Reference)

Bridge #282, the Lyon Brook Bridge

During the year 1869 the construction of the New York & Oswego Midland Rail Road began in earnest. Bridging the Lyon Brook ravine was a difficult engineering feat that had to be completed to begin train operations of the Midland.

By the middle of 1869 workmen began arriving to commence work on what would become the engineering landmark of the Oswego Midland's Northern Division, the bridge over the Lyon Brook. This bridge, located between Norwich and Oxford, was planned to be 830 feet long, and its center 100 foot span was to be located 155 feet above the namesake brook. The abutments needed to hold the iron superstructure were put up, utilizing, in part, flagstones from the nearby John Shattuck farm. Caps were put on top of the abutments and everything was in order to begin erecting the ironwork. Contractors Smith, Latrobe, & Co. began receiving the bridge iron members from the Phoenix Iron Works in Philadelphia. With Mr. Latrobe personally supervising the placement of the iron members, the lofty crossing of the ravine began. During this period of construction a crowd of onlookers seemed to always be present. These curiosity seekers marveled at how the workmen were able to work so high in the air.

On October 15th, the 100 foot Fink Truss section was put in place directly over the ravine. Through November and on into December the work continued. Finally, just two days before the Christmas of 1869, with the trackwork installed atop the iron bridge, the first train was sent onto the completed bridge to test the span. Though there were nearly 2000 people in attendance to witness the trial of the bridge, not a sound was uttered until the train (consisting of a locomotive, 6 flat cars, and 2 passenger coaches) reached terra firma after crossing the span. The crowd cheered, the locomotive whistle was blown, all done to announce that man had overcome a natural barrier.

Even before regular train operations to Sidney Plains began (on June 15, 1870), passenger excursion trains were regularly run to the bridge for people from far and near to view the completed wonder for themselves. It appears that the first excursion to the completed bridge was run on New Years Day of 1870. As reported in the Norwich Chenango Union newspaper, "upon arriving at the north end of the bridge, the train stopped, and most of those onboard walked out onto the bridge which was fitted up with substantial railings and a plank walk on both sides. Looking down from the dizzy height, at the seemingly frail supports, one can hardly believe it possible that the bridge can bear the heavy burdens which must cross it. After viewing the structure, the company repaired to the train which then slowly ran onto the bridge as far as the center where it halted a few minutes before starting back to Norwich". On a subsequent trip run on May 5, 1870, some passengers were not content to view the bridge from just the top. "We clamered down the banks of the Lyon Brook and into the chasm over which the celebrated structure stood, its iron network resembling more a horizontal cobweb than the firm and substantial work it is. Its height cannot be appreciated by riding over it. One must go down and then look up to be conscious of its altitude. We advise everyone not to miss a sight of it from the level of the water." Whether viewed from above or below, the Lyon Brook Bridge was an impressive structure that by itself alone brought much needed revenue to the fledgling Midland Railroad. One early excursion is pictured in Figure 4.23 showing the train just getting fully onto the bridge from the north end. The locomotive is #4, the "Delaware" with early coaches coupled to its tender.

With the opening of the railroad to Sidney Plains, the Lyon Brook Bridge was put to the test on numerous occasions daily and served the Midland well despite the railroad's financial difficulties. When the railroad was sold to a group of investors and became the New York, Ontario & Western Railway the bridge continued its use. Ten years after the O&W took over the operation of the old Midland, President Fowler officially opened the railroad's new Scranton Branch, on June 30, 1890. This new line, which tapped the lucrative coal fields of the Lackawanna Valley, brought about the O&W's entrance as a coal hauler. To more efficiently handle this increased tonnage, the O&W set about making the line less up and down hill and also completely rebuilt the Lyon Brook Bridge.

Contracts for the rebuilding of the bridge were made during February of 1894. The masonry for the new piers was begun in May, and during June the erection of the new steelwork commenced. The new structure contained eight plate girder tower spans of 30 feet each, three spans of 80 feet, four spans of 60 feet, and two spans of 50 feet, a total length of 820 feet. On August 15, 1894 the "new" bridge #282 was completed at a cost of $35,342.51. Surprisingly, all the work was done without an interruption of train service! Now, heavily loaded coal trains, powered by more modern locomotives, were able to speed towards their destination of either Utica, Rome, or Oswego untethered to the weight restrictions of the original bridge.

As expected, this structure was a very important segment of the railroad. So important that during the years of World War II the bridge was guarded 24 hours a day to protect it from any form of sabotage. An observation post was established on the south side of the bridge, and a small gable roofed shanty put in place to house the watchmen.

The rebuilt Lyon Brook bridge served the O&W well until the 20th century but was finally abandoned on March 29, 1957. For another eight years the structure stood over the Lyon Brook, but the structure was finally dismantled leaving nothing but the supporting piers, beginning in March, 1966. A marker has been placed along East River Road (just south of Polkville, N.Y.), marking the location of this historic structure.

This famous bridge is featured in the March 1999 edition of Model Railroader magazine. Original drawings of the bridge follow showing the bridge in cross section, the truss section and plate girder section.

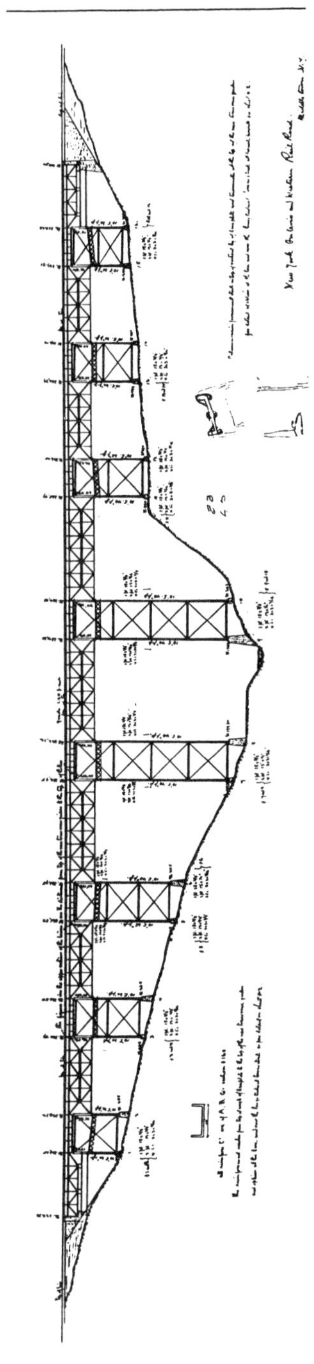

Figure 4.24 Cross Section of the Lyons Brook Bridge
Reprinted with permission of Ontario& Western Historical Railway Society

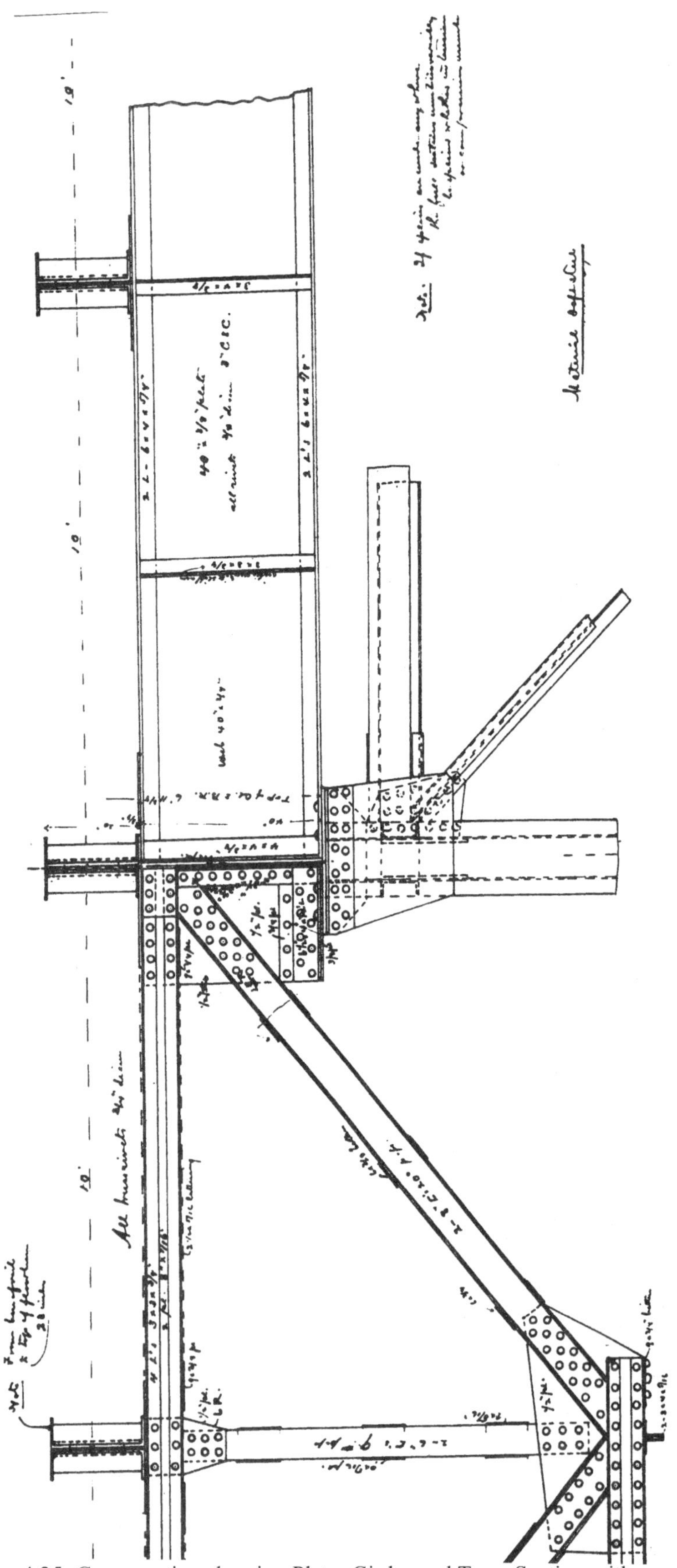

Figure 4.25 Cross section showing Plate Girder and Truss Section with connections
Reprinted with permission Ontario & Western Historical Railway Society

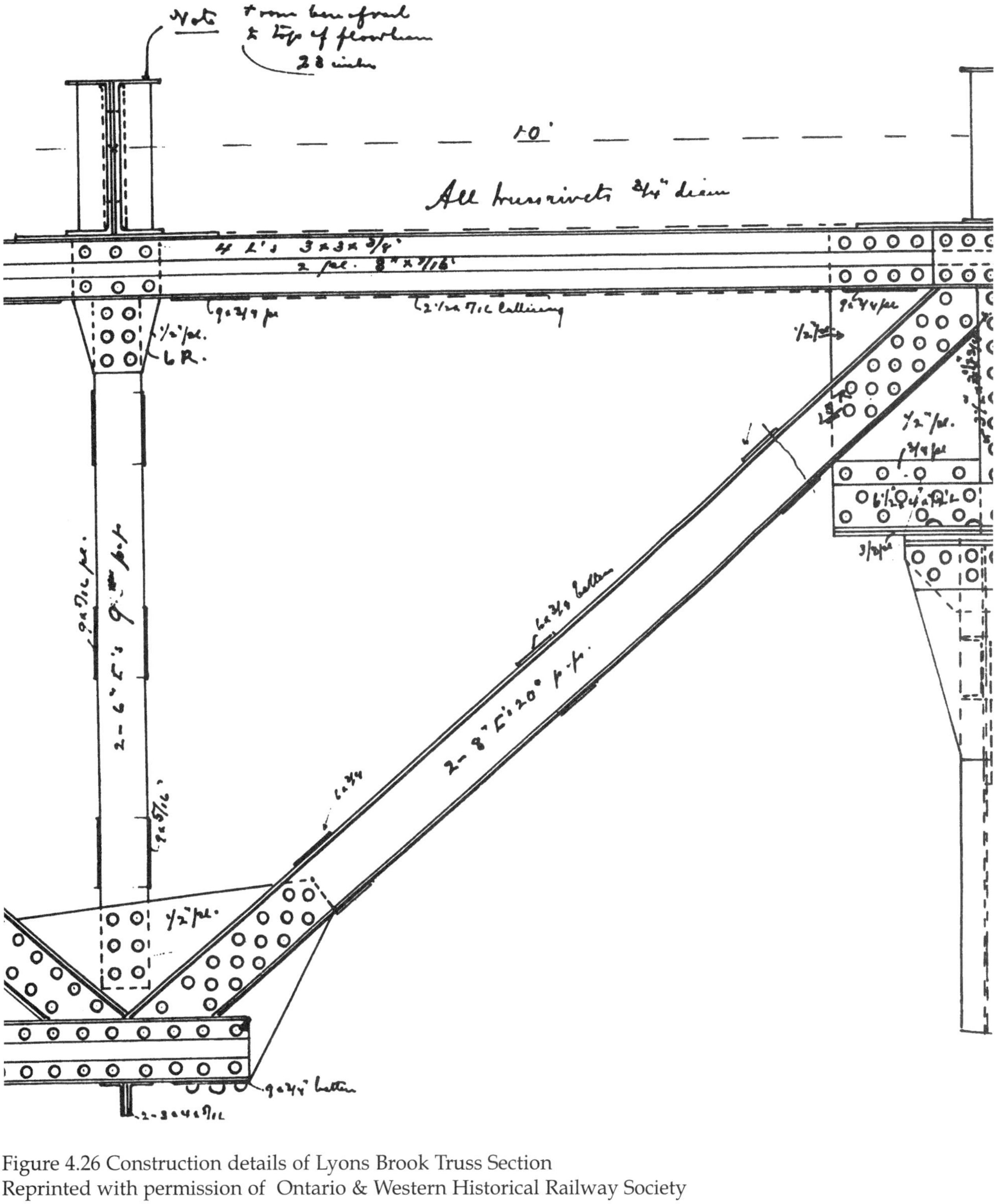

Figure 4.26 Construction details of Lyons Brook Truss Section
Reprinted with permission of Ontario & Western Historical Railway Society

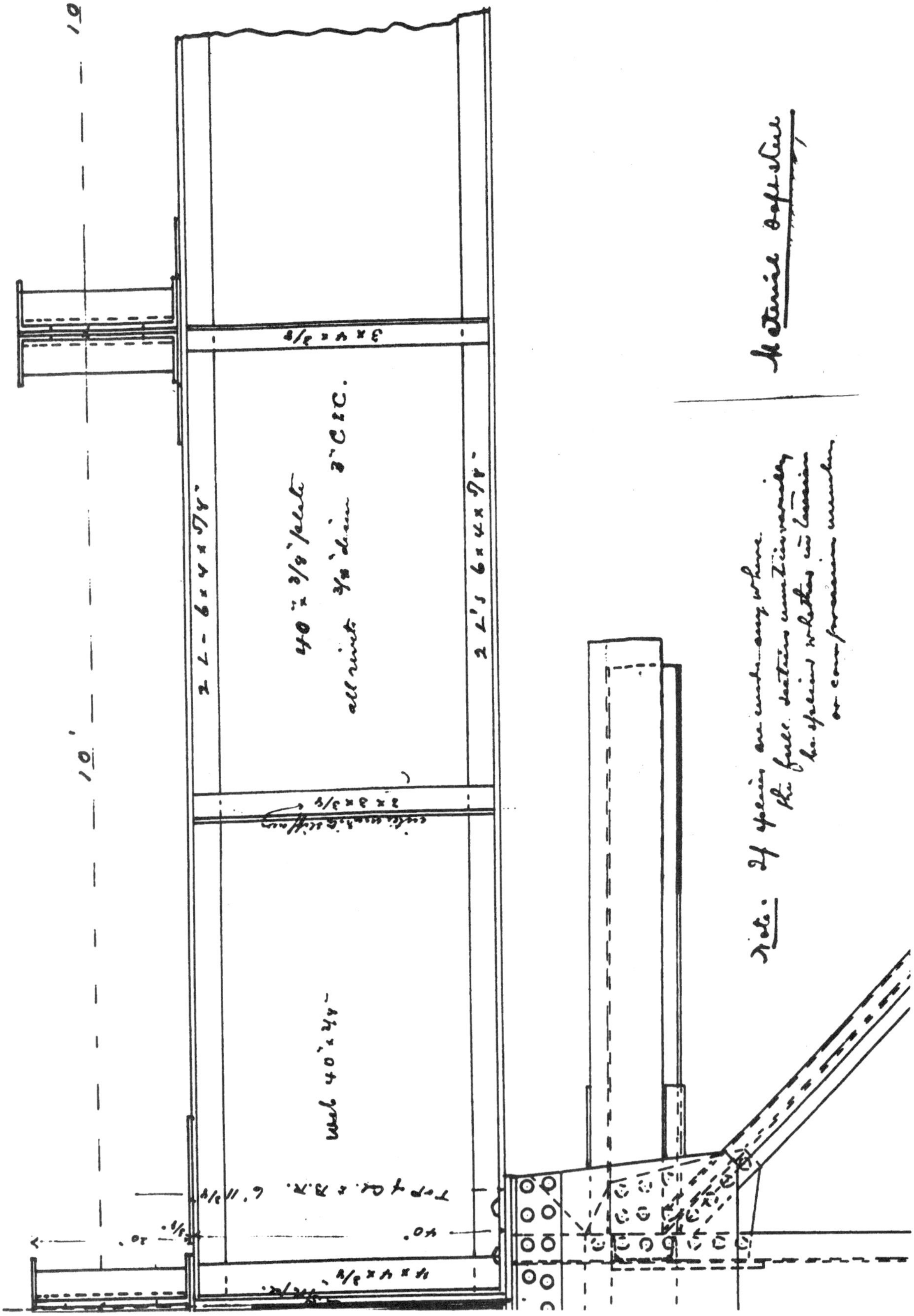

Figure 4.27 Construction details of Lyons Brook Plate Girder Section
Reprinted with permission of Ontario & Western Historical Railway Society

4.7 Bridge Supports - Trestles and Piers

Obviously, bridge trusses need to be constructed on sturdy supports to carry the loads on the bridge. Bridge piers are needed in rivers as a foundation for a bridge. Figure 4.29 shows the piers designed for the Los Animas River crossing with bridge 452-A, circa 1927. From such piers, trestles were normally constructed to the height needed to support the truss; figure 4.28 shows a standard pile trestle, circa 1907. These drawings are included as they may be useful to modelers developing valley crossings.

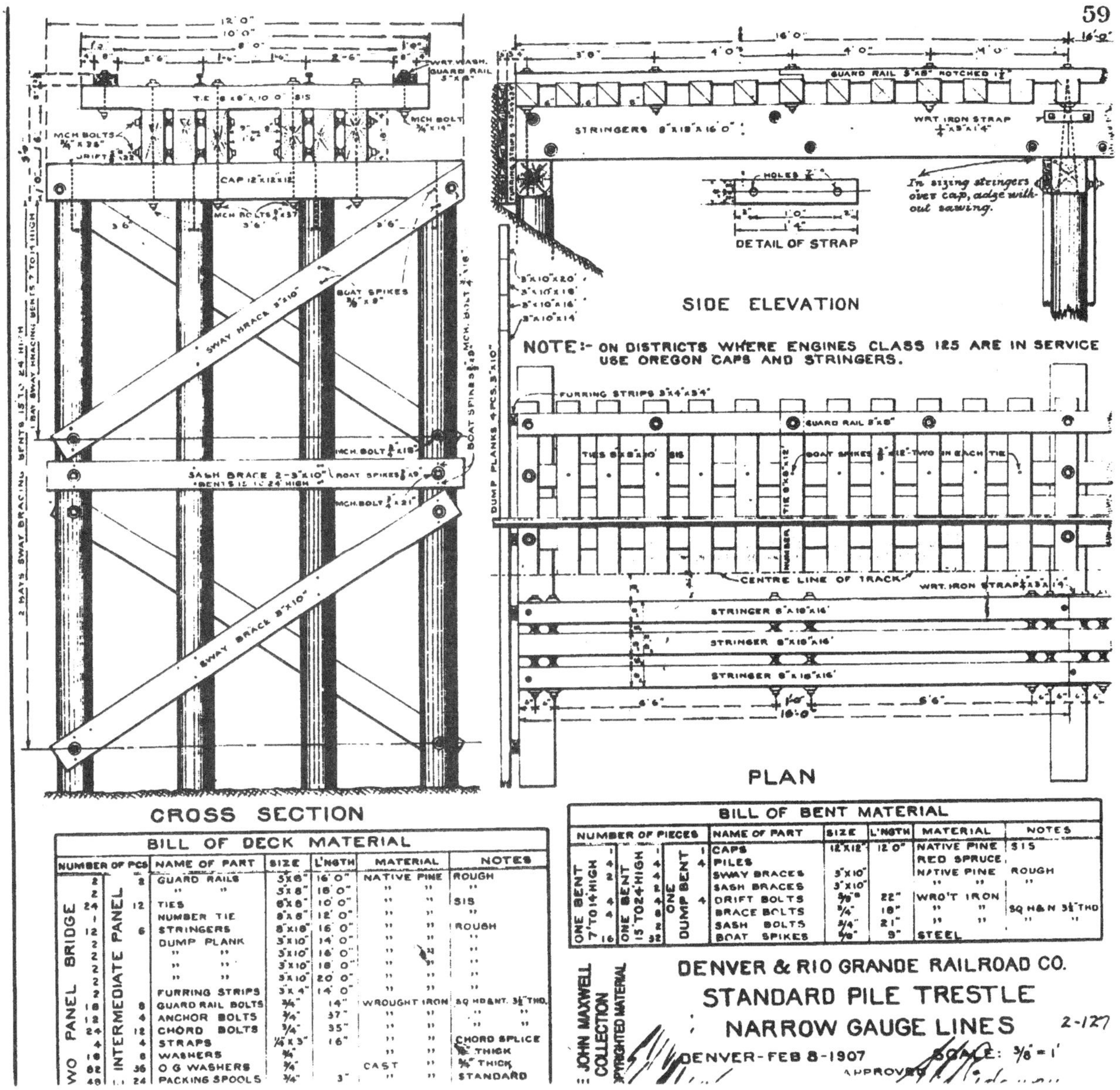

Figure 4.28 - Standard Pile Trestle for Narrow Gage Lines
Reprinted with permission: John Maxwell Collection, 1055 Jessica, Fernley, NV 89408

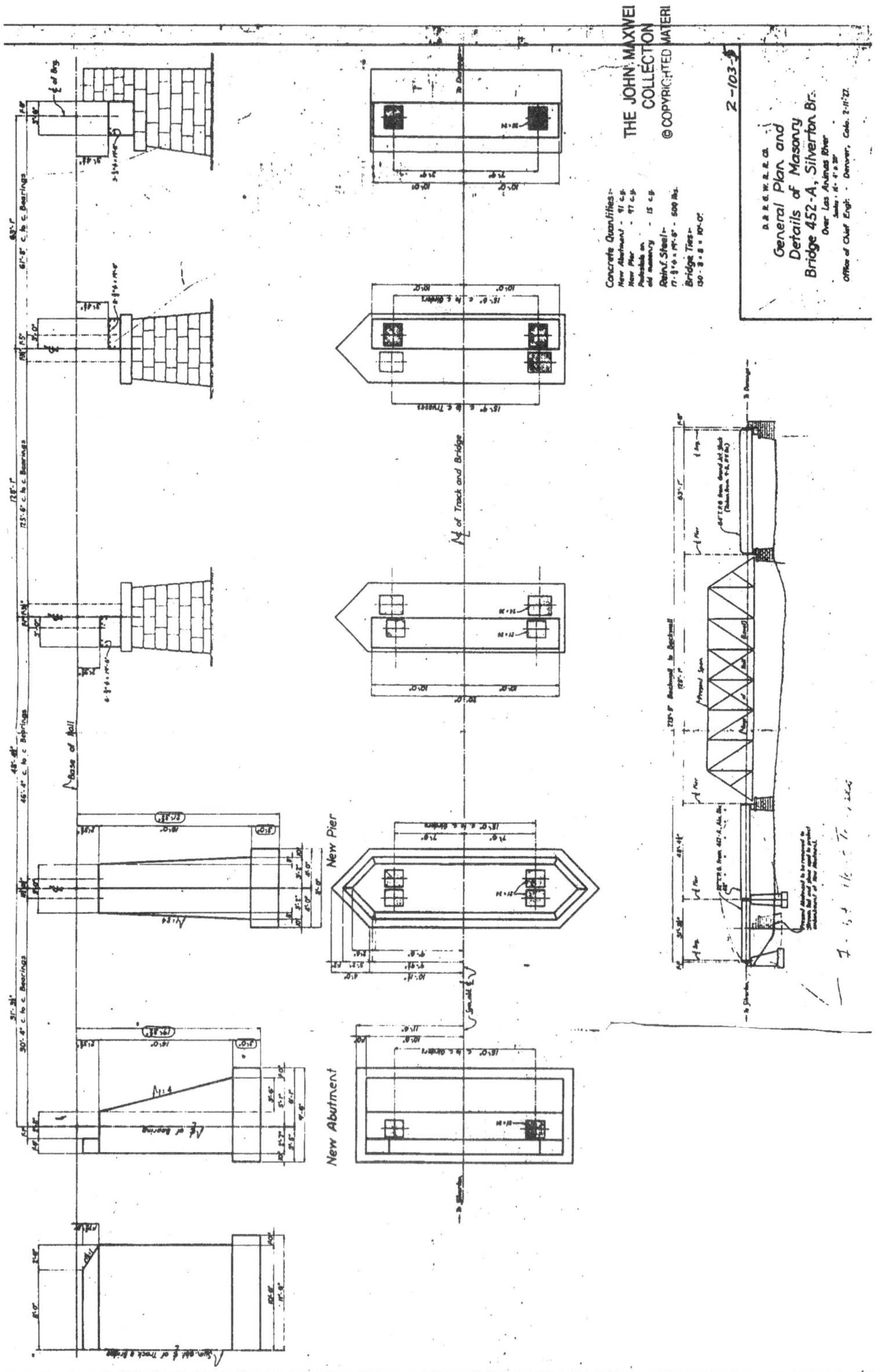

Figure 4.29 - Bridge Piers for Bridge 452
Reprinted with permission: John Maxwell Collection, 1055 Jessica, Fernley, NV 89408

Chapter 5 - Geometric Scale - Making Models

A lot of modeling is done in people's work, some modeling is done as a hobby or for fun. Architects make models of the structures they are designing to see how the final structures such as houses and other buildings will look. Movie makers make models to film events in the model studio that are difficult or impossible to film in real life. There are famous model studios which have made science fiction movies from models that look real. Monster movies frequently are made with models, or historic recreations can be made with models.

5.1 Calculating a scale

The scale of a model is the size of the model compared to the size of the real object being modeled. Suppose you want to know the scale of a model car. Measure the distance from the center of the front wheel to the center of the back wheel. This distance is the wheel base, for example 110" for a 4 door sedan. If a typical model car has a wheel base of 5.5" then the scale is 5.5"/110" . Usually this number is expressed with a "1" in the numerator, so this fraction would be reduced and a whole number used in the denominator to 1/20. This scale number means that the model will be 1/20 of the real car, or the real car will be 20 times larger than the model.

5.2 Scales from a ruler called a scale

Hobbyists use scales all the time to make models, small versions of things as they exist in real life. One of the most popular scale model makers are Model Railroader's. Other modelers make cars, military figures, or equipment from many different periods in history.

Models are made to a scale. We can choose any scale we want, but there are tools and premade pieces we can buy which are already sized to certain scales. For this reason, choosing a standard scale for a model is beneficial. Scales for different hobbies and uses follow:

Typical Model Train Scales

G	1/22.5
1	1/32
O	1/45
S	1/64
HO	1/87
TT	1/120
N	1/160
Z	1/220

Typical Architect's Scales
1/8" = 1'
1/4" = 1'
1/2" = 1'

Typical Civil Engineering Scales
1" = 50'
1" = 20'
1" = 10'

The figures which follow demonstrate how scale affects the final model size. The figure is approximate and should not be used for actual model construction but could be used for planning purposes. Most hobby stores, art supply stores, and many catalogs sell single scale rulers which help measure feet in the model scale, or 3 sided rulers, called scales, which measure scale distances as if they are feet. For example, if a model was developed in HO scale, 87 feet in real life would cover a distance of 1 foot in a model. We could make a ruler, like the charts which follow - see HO on page 2, that would have a mark at 1 foot and would be labeled as 87. A distance of 43.5 feet in real life would be 6 inches in the model. But what would be more convenient would be a ruler that when it said "10", that would mean that the model represents 10 feet in real life, (but the mark on our ruler would be at 10/87 feet or 1.4"). In the following figure , the mark that says 10 feet for a HO model is drawn at 1.4". A scale makes converting real life distances into model distances easy. If you do a lot of modeling, you will need a scale!

Larger scales -1 foot rulers

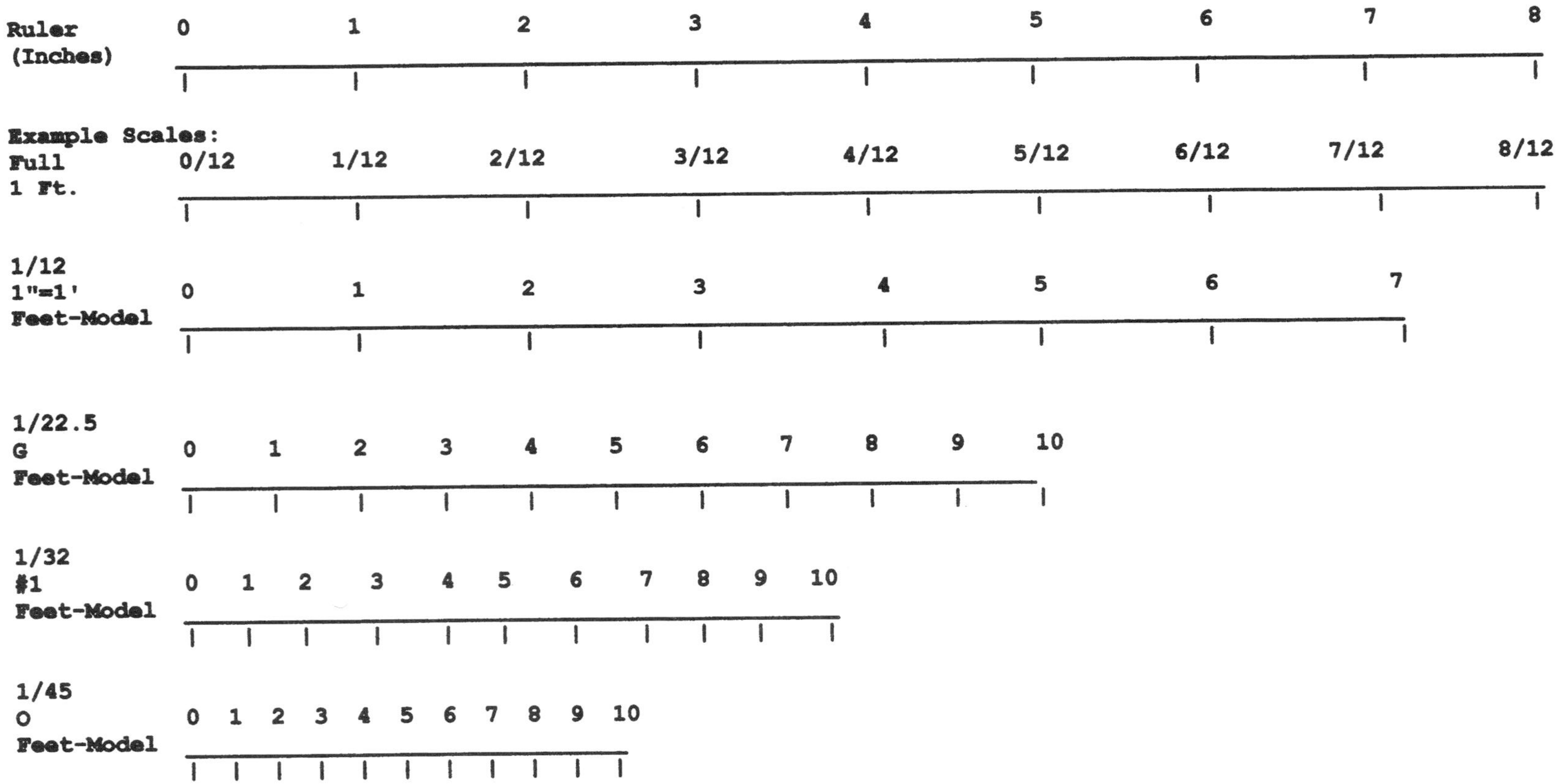

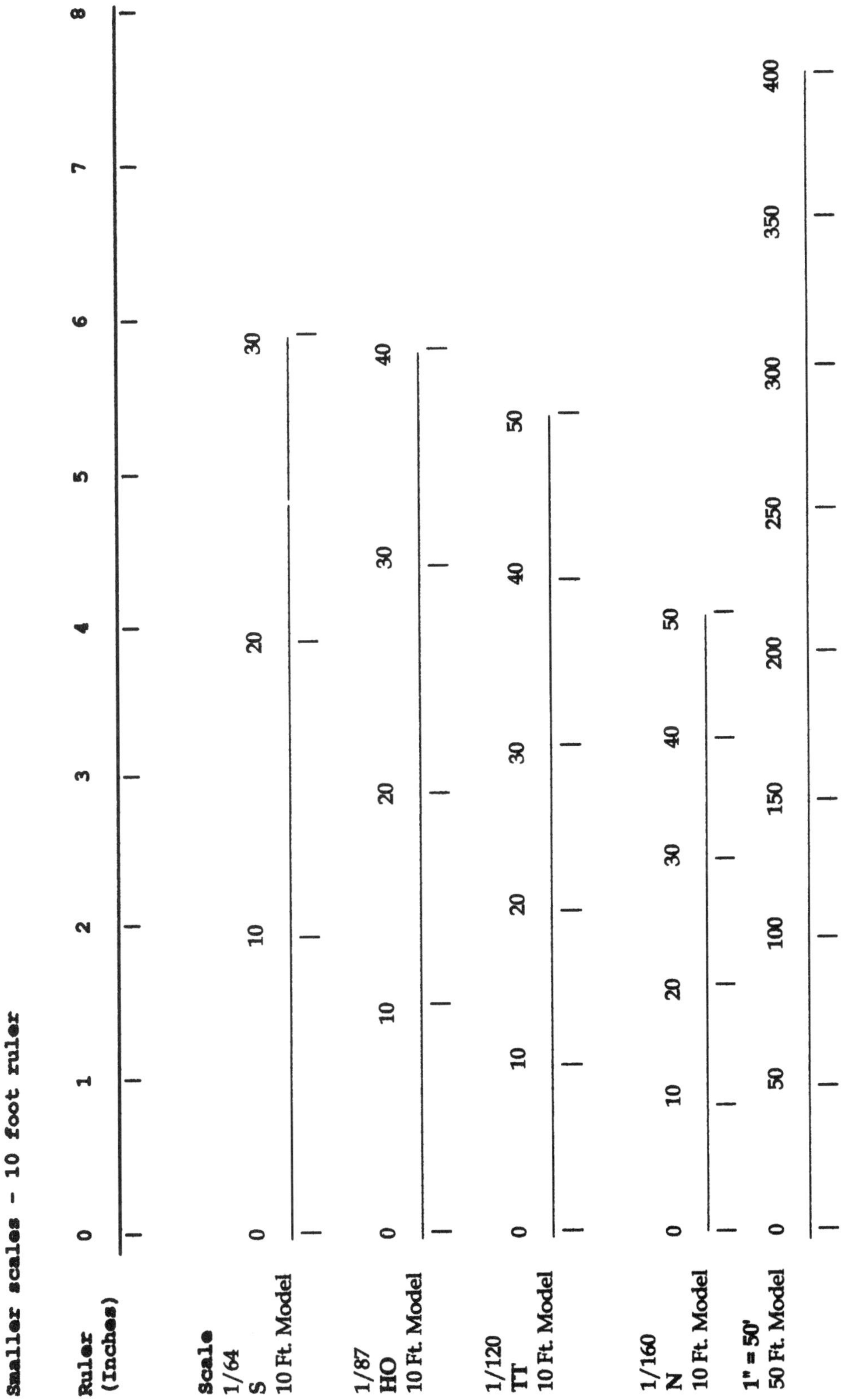
Smaller scales - 10 foot ruler
Ruler (Inches)
0 1 2 3 4 5 6 7 8
Scale
1/64
S
10 Ft. Model
0 10 20 30
1/87
HO
10 Ft. Model
0 10 20 30 40
1/120
TT
10 Ft. Model
0 10 20 30 40 50
1/160
N
10 Ft. Model
0 10 20 30 40 50
1" = 50'
50 Ft. Model
0 50 100 150 200 250 300 350 400

5.3 Standard sizes and shapes of construction materials

Lumber that we buy in the lumber store comes in standard sizes. Many people are familiar with a 2x4 (pronounced 2 by 4) which is rough sawn to a 2"x4" size. When planed smooth the 2x4 is smaller than 2" x 4" but is still referred to as a 2x4. Lumber comes in many standard sizes like 4x4, 2x6, 2x8, 2x10, 2x12, 1x4, 6x6. Each size is available in various lengths like 8 feet, 10 feet, 12 feet, 16 feet, etc.

Steel is made for construction at the steel plant and also comes in standard sizes and shapes. Shapes are referred to as I, W, H, L, C, Z and some others. I and W sections are the most common sections used as beams. Beams are horizontal members used to build floors and other structures that are required to span a distance. A W section is made from three flat pieces with only slight rounding from the vertical line. The I section was introduced earlier, and was shaped by the machines available at the time. Both sections look like a capital I with the I section being thicker. The W section has replaced the I section in popularity of use for most cases.

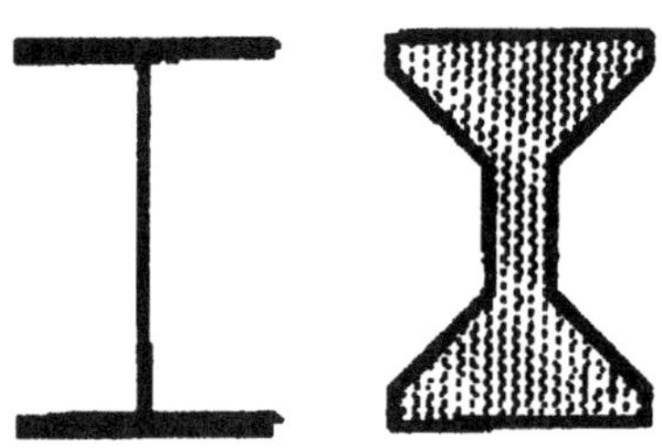

Figure 5.2 Wide Flange (W) and I (S) sections
Courtesy, University of Florida

An H section is much like a W section but is nearly square and often used as a column. A C section is a channel with 2 short legs and looks like a flat capital C. An L section is L shaped and is often used as a brace, and a Z section is Z shaped and often used to construct roofs.

Steel uses 2 dimensions in its name also. The first number is the height of the section in inches. The second number is the weight in pounds of the section per foot. For example, a W14x38 would be a W section, 14" tall and weighing 38 pounds per foot. Generally the larger the weight per foot, the more load the section will carry.

5.4 Design properties of construction sections

When designing a structure, one of the most commonly needed properties for members in the structure is the cross sectional area. For a rectangular section it is simply the width times the height. For other standard members, which are more complicated, tables are available for the area. Examples of these data are shown in Table 5.5.

In addition to cross sectional area, there are two other parameters which are important when we consider bending (covered in the next chapter). The first is called, "c", the distance from the center to the outside of the beam. Let's say we are studying the loads on a floor joist(used as a beam) in the floor of a house. This beam is probably a 2" x 6" piece of wood. The distance, c, is measured from the center of the board to the outside of the board, in this case 3". For rectangles, c, will always be ½ of the height. In fact, any shape which is symmetrical, that is, the same above and below the center line, will always have a value of c which is half of the height.

Other types of sections such as "T" sections obviously are not the same above and below the point of half height. Imagine that a capital "T" has to be balanced on your finger. Your finger will need to be closer to the big part on top, that is the centroid. The distance from this point to the top will be different from the distance from this point and down.

A "C" section is symmetric from top to bottom but is very different from left to right; one side has the metal for the "C" and the other side has none. You can't balance a capital "C" on your finger because where the C would balance there is no place to put your finger. It is probably clear, though, that the point would be closer to the left than to the right.

For all sections, the balancing point for the section is called the centroid, the geometric center. Typical values of the centroid for "C" and other sections are shown in Table 5.5.

The other important property we need to know for bending calculations is referred to as the Moment of Inertia, I. There's no easy way to explain Moment of Inertia. Generally it is the idea that an area farther away from the centroid resists bending better than the same area near the centroid. That explains why steel is shaped with a larger area placed away from the centroid, not flat like a board. You probably knew this, or could imagine it already. If you stood on a board, a 2"x6", would it deflect more if the board was flat, or if the board was standing up? (If you try this be careful, the board can tip.)

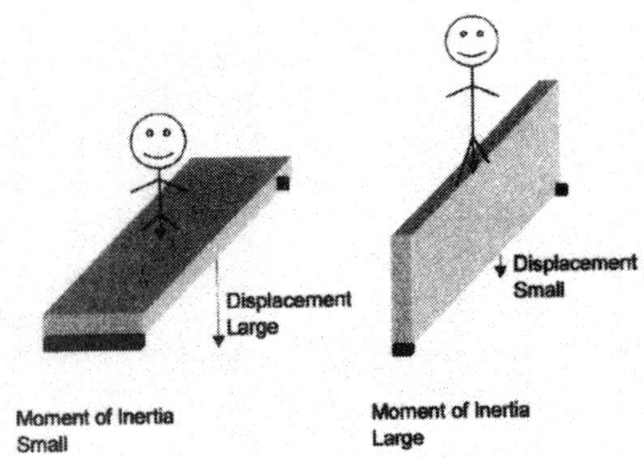

Figure 5.3 Board positioned in Minimum and Maximum Moment of Inertia

It would deflect less if the 2"x6" was upright. The Moment of Inertia is why; the board standing upright has a larger Moment of Inertia.. The equation to find Moment of Inertia for rectangular sections is:

$$I = 1/12\, wh^3$$

Where:

w= the width of the beam
h= the height of the beam

For nonrectangular sections it takes a few steps to calculate, but, tables are available for the engineer to look up the values. The larger Moment of Inertia is tabulated as I_{max} while the minimum is shown as I_{min}. The values of c_{max} and c_{min} correspond to these directions of the Moment of Inertia.

5.5 Hobby Modeler's Design Tables

Hobbyists need to be able to produce models with the correct overall dimensions for their model and correct dimensions for each piece used to construct the model. Members and elements that are too large or too small do not create the correct appearance. Some simple tables follow to aid the model designer in sizing structural members appropriately. Examples of use of these tables are in the workbook section which follows as the second part of this book.

Modeling of rods and tension members in trusses requires the conversion of diameters to scale proportions. Table 5.1 shows selected diameters (in inches) and the appropriate diameter (in inches) for different model scales. Table 5.2 aids in the conversion from wire gage to inches represented in a model scale.

Early trusses such as the Lattice Truss and the Howe Truss were constructed from lumber. The Lattice Truss had the advantage of using standard dimension lumber while the Howe Truss used large timbers. Table 5.3 provides a conversion of various common lumber sizes to various scales. Balsa wood and basswood are available in some hobby stores already sized for building structures using common railroad modeling scales.

Steel truss sections use common steel shapes. Standard sections are available in plastic from hobby supply stores for models. Table 5.4 shows typical hobby store product sizes with the height of the section listed as H, the width of the section listed as W, and the thickness of the plastic listed as T. Some example sections are listed under the scale headings which would be representative of the members used on a bridge if the model section was used. These are approximate for geometric modeling of the size of the member. The thickness conversion would not be correct in the modeled members as the thicknesses are generally too large in comparison to the field members to be scaled. Actual section properties are shown in Table 5.5 for the example sections in Table 5.4.

Table 5.1 Diameter to scale conversion

Diameter (In.)	G 1/22.5	O 1/48	HO 1/87	N 1/160
1/8	0.005	0.003	0.001	0.0008
1/4	0.010	0.005	0.003	0.0016
1/2	0.020	0.010	0.006	0.0031
3/4	0.030	0.015	0.009	0.0047
1	0.042	0.021	0.011	0.0063
2	0.084	0.042	0.022	0.0125

Table 5.2 Wire Gage to scale inches

Wire Gage	G 1/22.5	O 1/48	HO 1/87	N 1/160
5	4.10	8.75		
10	2.34	5.00	9.00	
15	1.29	2.75	5.00	9.00
16	1.16	2.48	4.50	8.00
18	0.94	2.00	3.50	6.50
20	0.70	1.50	2.72	5.00
22	0.60	1.25	2.18	4.00
24	0.47	1.00	1.78	3.25
26	0.35	0.75	1.36	2.50
30	0.23	0.50	0.91	1.66
34	0.13	0.28	0.50	1.00
40	0.06	0.14	0.25	0.50

Table 5.3 Common lumber sizes to scale sizes

Lumber Size	G 1/24	O 1/48	HO 1/87	N 1/160
2" x 4"	0.08" x 0.16"	0.04" x 0.08	0.02" x 0.09"	0.013"x 0.025"
2" x 6"	0.08" x 0.25"	0.04" x 0.13"	0.02" x 0.07"	0.013"x 0.038"
2" x 8"	0.08" x 0.33"	0.04" x 0.17"	0.02" x 0.09"	0.013"x 0.050"
4" x 4"	0.17" x 0.17"	0.08" 0.08"	0.05" x 0.05"	0.025"x 0.025"
6" x 6"	1/4" x 1/4" 0.25" x 9.25"	1/8" x 1/8" 0.13" x 0.13"	0.07" x 0.07"	0.038"x 0.038"
6" x 8"	0.25" x 0.33"	0.13" x 0.17"	0.07" x 0.09"	0.038"x 0.050"
6" x 10"	0.25" x 0.42"	0.13" x 0.21"	0.07" x 0.11"	0.038"x 0.063"
8" x 8"	0.33" x 0.33"	0.17" x 0.17"	0.09" x 0.09"	0.050"x 0.050"
8" x 10"	0.33" x 0.42"	0.17" x 0.21"	0.09" x 0.11"	0.050"x 0.063"
10" x 10"	0.42" x 0.42"	0.21" x 0.21"	0.11" x 0.11"	0.063"x 0.063"
10" x 12"	0.42" x 0.5"	0.21" x 0.25"	0.11" x 0.14"	0.063" x 0.75"
12" x 12"	1/2" x 1/2" 0.5" x 0.5"	1/4" x 1/4" 0.25" x 0.25"	1/8" x 1/8" 0.14" x 0.14"	0.075"x 0.075"
12" x 14"	0.5" x 0.58"	0.25" x 0.33"	0.14" x 0.16"	0.075"x 0.087"
12" x 16"	0.5" x 0.66"	0.25" x 0.37"	0.14" x 0.18"	0.075"x 0.1"
16" x 16"	0.66" x 0.66"	0.37" x 0.37"	0.18" x 0.18"	0.1"x 0.1"

Table 5.4 Typical Modeling supply items converted to Approximate Steel Section sizes

I-Beams (S sections)

Product H (h)	Product W (b)	Product T (t)	G 1/22.5	O 1/48	HO 1/87	N 1/160
1/16	1/32	0.015				S10*35
3/32	3/64	0.02			S8x18.4	
1/8	1/16	0.025		S6x12.5	S10x35	
5/32	5/64	0.03		S7x20)	W14*38	
3/16	3/32	0.035			W16x58	W30x210
1/4	1/8	0.035	W6x8.5	S12x35		
5/16	5/32	0.045	S7*20		W27x145	
3/8	3/16	0.045		W18x85		
7/16	7/32	0.055	W10x21			
1/2	1/4	0.060	S12x50	W24x100		
9/16	9/32	0.060	W14x38	W27x145		
5/8	5/16	0.060	S15x50	W30x190		

C Sections (Channels)

Product H	Product W	Product T	G 1/22.5	O 1/48	HO	N
1/16	0.035	0.015		C3x6		
1/8	0.050	0.025		C6x13		
3/16	1/16	0.035	C4x7.25			
1/4	5/64	0.035	C6x8.2	C12x30		
5/16	3/32	0.045	C8x18.75			

H sections

Product H	Product W	Product T	G 1/22.5	O 1/48	HO 1/87	N 1/160
1/16	1/16	0.02				HP 10x57
3/32	3/32	0.035			HP 8x36	
5/32	5/32	0.030		HP 8x36	HP 14x117	
1/4	1/4	0.04		HP 12x74		
5/16	5/16	0.045	HP 8x36			

Note: Product dimensions taken from Plastruct Catalog. 1020 S. Wallace Place, City of Industry, CA 91748. Chosen sections are approximate for height and width, thickness is not to scale.

Table 5.5 - Properties of Sections

Section	h	b	t	Area	c_{max}	c_{min}	I_{max}	I_{min}
S 6x12.5	6.00	3.33	0.232	3.67	3.00	1.66	22.1	1.82
S 7x20	7.00	3.86	0.450	5.88	3.50	1.93	42.4	3.17
S 8x18.4	8.00	4.00	0.271	5.41	4.00	2.0	57.6	3.73
S 10x35	10.00	4.94	0.594	10.30	5.00	2.47	147.0	8.36
S 12x35	12.00	5.08	0.428	10.30	6.00	2.54	229.0	9.87
S 12x50	12.00	5.47	0.687	14.70	6.00	2.74	305.0	15.7
S 15x50	15.00	5.64	0.550	14.70	7.50	2.82	486.0	15.7
W 6x8.5	5.83	3.94	0.170	2.51	2.92	1.97	14.8	1.98
W 10x21	9.90	5.75	0.240	6.20	4.95	2.88	107	10.8
W 14x38	14.12	6.78	0.313	11.20	7.06	3.39	386	26.6
W 16x58	15.86	8.46	0.407	17.10	7.93	4.23	748	65.3
W 18x35	17.71	6.00	0.298	10.30	5.15	3.00	513	15.5
W 24x100	24.00	12.00	0.468	29.50	12.00	6.00	3000	223
W 27x145	26.88	13.97	0.600	42.70	13.44	6.99	5430	443
W 30x190	30.12	15.04	0.710	56.00	15.06	7.52	8850	673
W 30x210	30.38	15.11	0.775	61.9	15.19	7.55	9890	757
C 3x6	3.00	1.60	0.356	1.76	1.50	0.45	2.07	0.31
C 4x7.25	4.00	1.72	0.321	2.13	2.000	0.86	4.59	0.43
C 6x8.2	6.00	1.92	0.200	2.40	3.00	0.51	13.1	0.692
C 6x13	6.00	2.157	0.437	3.83	3.00	0.51	17.4	1.05
C 8x18.75	8.00	2.527	0.487	5.51	4.00	0.57	44	1.98
C 12x30	12.00	3.170	0.510	8.82	6.00	0.674	162	5.14
HP 8x36	8.03	8.158	0.446	10.60	4.01	4.079	120	40.4
HP 10x57	10.01	10.224	0.564	16.80	5.00	5.112	295	101
HP 12x74	12.12	12.217	0.607	21.80	6.06	6.109	566	185
HP 14x117	14.23	14.885	0.805	34.40	7.15	7.443	1230	443

Note: Properties from American Institute of Steel Construction, Inc. 1970 (and calculated)

5.6 Making an HO Scale Model

Construction drawings, such as those in Chapter 4, can be used to produce models at any chosen scale. An example is shown in Figure 5.4. This figure shows a typical layout and is to HO scale. Hand written notes from the original drawing (but too small in the reproduction) are also shown in the figure. The intent of this drawing is not to produce a detailed scaled reproduction of the original drawing but rather to provide an outline which can be used to cut wood pieces for the construction of the bridge. The process to produce this diagram is fun. First the dimensions from the original, see Figure 4.14, are converted to scale using a simple spreadsheet program to make the table shown in Table 5.6. Such a table can be transferred to a piece of graph paper or drawn with a scale to produce the outline drawing in Figure 5.4. For assistance in modeling, the cross section showing the track placement is shown in figure 4.15. Other details may be obtained by studying the original drawings.

The lower chord of the bridge contains a 2 foot extension (beyond the first cross braces for the truss) and then 7' 4" panels for a total length of 48 feet. The upper chord also contains a 2 foot extension, with a total length of 33' 5". In Figure 5.4, the lines used to represent the upper and lower chords are the inside edge of the timbers. Panel cross members forming the "X" in the Howe truss are recreated using dashed lines. Vertical rods are shown as solid vertical lines. Notes identify the size of members. This information should be sufficient to build a realistic model.

- 48 foot Standard Howe Truss		Prototype	Model
Description	Dimension	Distance-Ft.	Distance-In.
Lower Chord		0	0
Entrance	2'	2	0.2758621
Diagonal Panel	7' 4"	9.3333333	1.2873563
Panel 1	7" 4"	16.666667	2.2988506
Panel 2	7" 4"	24	3.3103448
Panel 3	7" 4"	31.333333	4.3218391
Panel 4	7" 4"	38.666667	5.3333333
Diagonal Panel	7" 4"	46	6.3448276
Exit	2'	48	6.6206897
Upper Chord			0
Space to beginning	7'4"	7.3333333	1.0114943
Overhang	2'	9.3333333	1.2873563
Panel 1	7' 4-1/4"	16.6875	2.3017241
Panel 2	7' 4-1/4"	24.041667	3.316092
Panel 3	7' 4-1/4"	31.395833	4.3304598
Panel 4	7' 4-1/4"	38.75	5.3448276
Overhang	2'	40.75	5.6206897
Height	10' 8"	10.666667	1.471264

Table 5.6 Calculation of model dimensions from a standard drawing
Note:
The lower chord of the truss extends 2 feet beyond the beginning the first panel to provide the entrance to the truss. This is followed by 6 panels of 7' 4" and another 2 foot extension for the exit to the truss. The total length of the bottom chord is 48 feet. Similarly, the top chord has a 2 foot extension at each end resulting in a total length of 33' 5" (See Figure 5.3)

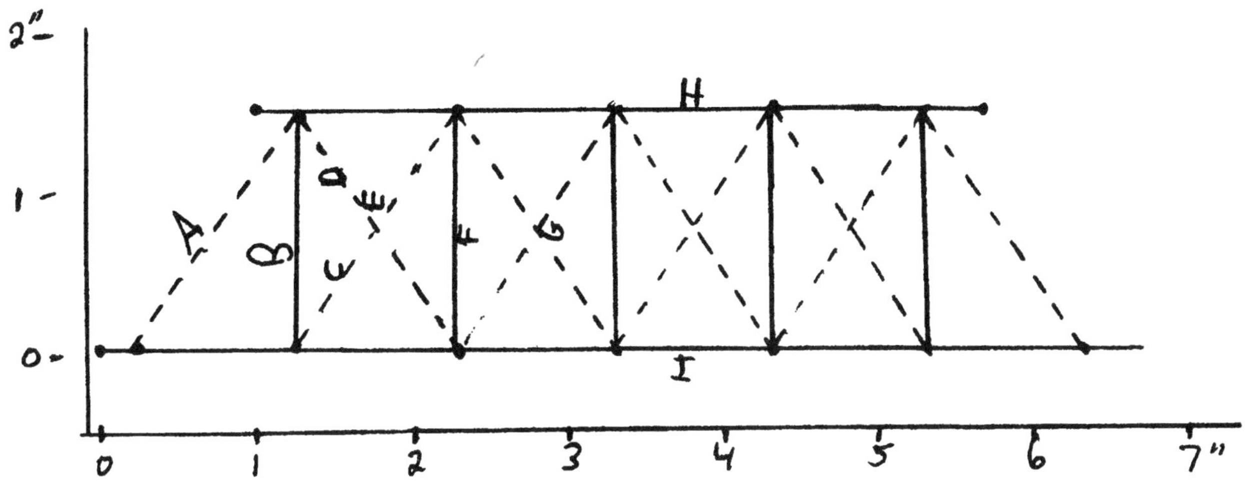

Figure 5.4 Layout drawing of a HO truss

Drawing Notes:	Converted to scale
A. 2 Pcs. 9 in x 10 in. 12' 3"/8' long	0.10" x 0.11" (Table 5.3) (cut to length from diagram)
B. 2 rods 1 3/8" diameter 2 rods 1 1/2"	0.014" diameter (Table 5.1)
C. 2 Pcs. 6" x 8" 12' 3-7/8" long	0.07" x 0.09" (Table 5.3)
D. 2 Pcs. 6" x 8" 12' 4" long	
E. 5/8" x 26" long bolt	0.005" Diameter (Table 5.1)
F. 2 rods 1-13/16"	0.020" Diameter (Table 5.1)
G. 2 Pcs. 8" x 8" 12'6" long	0.09" x 0.09" (Table 5.3)
H. 2 Pcs. 6" x 10"	0.07" x 0.11" (Table 5.3)
I. 2 Pcs. 6" x 12"	0.07" x 0.14" (Table 5.3)

Other notes:
Cross member between trusses 7" x 15" 18' long
Beam to support track base 6" x 12"

Table 6.1 Reference Table - Relationships of triangles

Length relationship $O^2 + A^2 = H^2$

Sum of angles = 180°

Sine = Opposite / Hypotenuse = O/H

Cosine = Adjacent/Hypotenuse = A/H

Tangent = Opposite / Adjacent = O/A

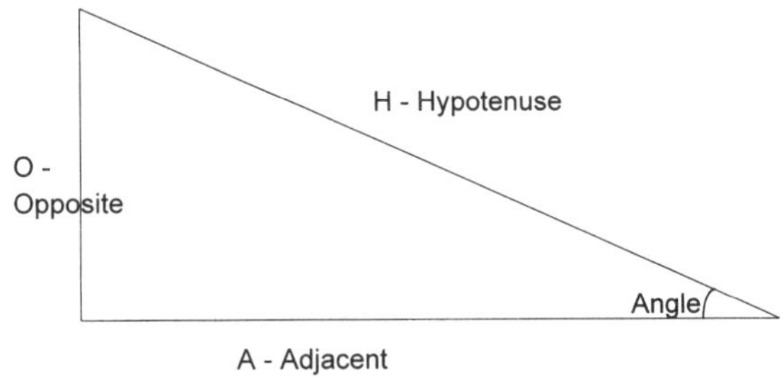

Some values of common angles

Angle	0	10	20	30	37	45	53	60	90
Sine	0.00	0.17	0.34	0.50	0.60	0.71	0.80	0.87	1.00
Cosine	1.00	0.98	0.94	0.87	0.80	0.71	0.60	0.50	0.00
Tangent	0.00	0.18	0.36	0.58	0.75	1.00	1.33	1.73	Inf.

Some famous triangles

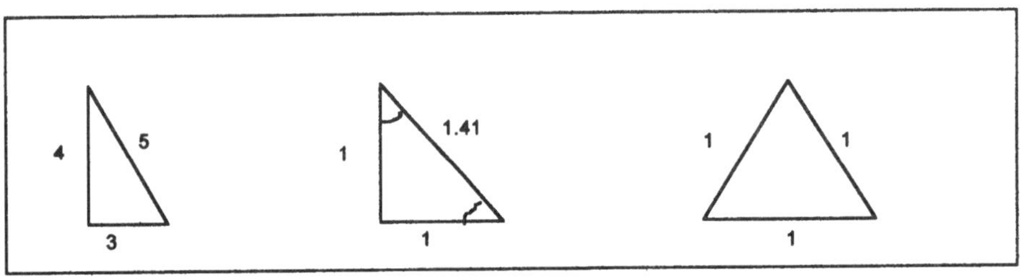

3-4-5 **45 degree** **Equilateral**

Chapter 6 - Analyzing Trusses

In Chapter 3, the triangle building block was introduced as a key principle in trusses. To analyze trusses the geometric relationships of triangles will be used extensively. First, geometry will be used to calculate the length of members and other needed parameters. A brief summary of some geometric relationships are shown in Table 6.1. Next, the principles of triangles can be used to analyze forces and their effect on structures. These two important principles are known as Vectors and Static Equilibrium.

6.1 The Directional Cosines

In the analysis of trusses the directional cosines are a quick and easy way to find unknown quantities. Directional cosines are simply the ratio of sides of the triangle. Similar triangles may be used in analysis. Similar triangle are triangles with similar ratios of sides. If we know that we have a triangle with sides 3-4-5, and we know that we have another triangle which is similar and has a hypotenuse of 60, then we can quickly find the other two sides as 60*(3/5) or 36, and 60*(4/5) or 48. This concept of directional cosines, and the use of similar triangles, makes analysis of trusses quick and fun.

6.2 Vectors

Forces can be represented by vectors. Vectors have a length and direction. The length of the vector represents the magnitude of the force. In short, vectors are force triangles and follow the same rules as triangles.

It is often convenient to convert a vector into its x and y components. An example of this is shown below.

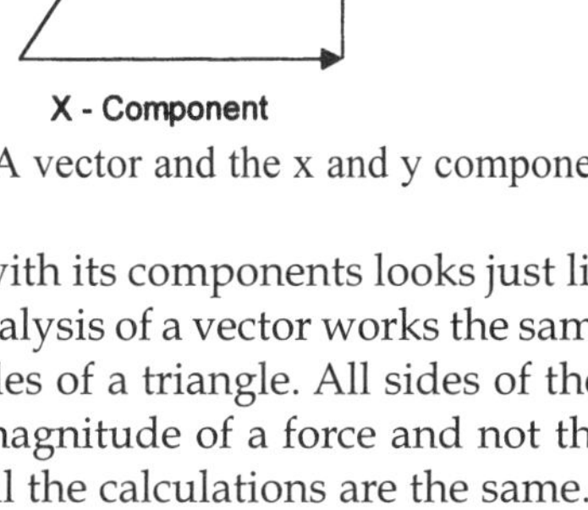

Figure 6.1 A vector and the x and y components

A vector with its components looks just like a triangle, and the analysis of a vector works the same as calculating the sides of a triangle. All sides of the triangle are now the magnitude of a force and not the length of a line, but all the calculations are the same.

Let's look at an example of the use of vectors. Imagine a person rowing a boat across a flowing stream. The rower can row the boat at a speed of 4 miles per hour. While the rower is rowing, the boat is being carried downstream by the flow of the stream which is 3 miles per hour. The stream is ½ mile across. How long will it take to row across the stream? If the rower rows straight north where will the boat be when it touches shore?

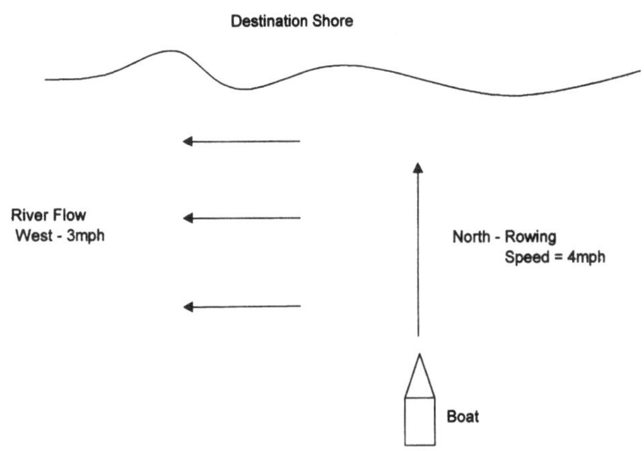

Figure 6.2 Example of vectors

The time it takes to row across is:

Time = Distance/Rate
Time = (½ mile)/(4 miles per hour)
= 1/8 hour or 7.5 minutes

As the boat is being rowed straight across the river, the current is carrying the boat downstream at the same time. The travel distance is:

Distance North = ½ mile (4/8 mile)
Distance West = 3 miles per hour x 1/8 hour = 3/8 miles

How far did the boat actually travel?

The boat travels the diagonal of a 3-4-5 triangle or 5/8 of a mile. (The boat traveled ½ mile north (4/8 mile) and 3/8 mile west thus the resultant is 5/8 mile)

Remember:
Vectors are force triangles.

6.3 Static Equilibrium

To analyze trusses the principles of Static Equilibrium are used. To be in static equilibrium the sum of the forces and moments acting on the structure (or member) must equal zero. If there are any left over forces or moments the unbalanced forces will cause an acceleration to the object and the unbalanced moments will cause rotation or spinning of the structure. Needless to say neither of these is desirable for a truss bridge. Stated in a formal way:

Static Equilibrium: The sum of all forces must equal zero for a structure to be in equilibrium.

There are three equations of equilibrium:

Sum $F_x = 0$ Sum of forces in x direction = 0

Sum $F_y = 0$ Sum of forces in y direction = 0

Sum M = 0 Sum of moments = 0

Unless otherwise indicated the common convention is x being horizontal (sideways on the page) and positive to the right, and y direction vertical (up and down on the page) and positive if up.

Moments are created by forces which act on a structure (or member) away from a support. If not countered these would cause the structure to rotate. Moments are calculated as:

$M = F \times d$

Where M = Moment
F = Force
d = distance

The three equations of equilibrium allow solutions of up to three unknowns at locations within a plane truss.

6.4 Boundary Conditions

Structures are attached to the earth with connections. These connections are formally called, boundary conditions. For analysis of trusses two boundary conditions are defined, as shown in Figure 6.3:

1. The pin, and, 2. The roller.

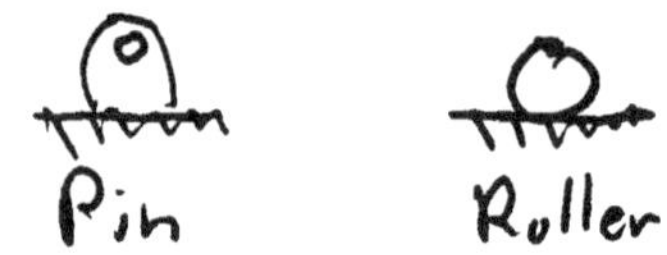

Figure 6.3 - Pin and Roller

(Note that in this chapter simple hand drawn sketches are used. This is done to encourage readers to try some problems themselves and learn how to draw and visualize problems with their own simple diagrams. It's fun.)

The pin represents a bracket which will not let the structure move up or down, left or right. The structure however, can rotate around the pin. Since the pin cannot resist moments, the Moment at the pin is zero, Sum M = 0. This is a point which can be used to sum moments equal to zero in the equations of static equilibrium. Thus all three equations of equilibrium may be used at a pin: The Sum of Moments = 0, The sum of forces = 0 (x and y). Note that there can be F_x and F_y forces in the x and y direction created by loads on the structure and held by the pin.

The roller will not let the structure move up or down but will allow movement left and right and will allow rotation. Note that the picture is a little misleading since it looks like a load could lift the structure off of the roller. The boundary condition is more like a roller in a slot or track so it can resist forces in both the up and down directions. Since a roller can resist only forces in the up and down direction (y direction) there is only one equation of equilibrium available.

For many bridges (but not all) it is typical to be supported at each end, one end having a pin (fixed end) and the other end having a roller (expansion end). This is called a simply supported structure. The roller (or expansion end) allows the structure to expand or contract with changes in temperature without adding stresses in the members of the structure.

6.5 Free body diagrams

To be able to "see" the forces in a structure we need to create a free body diagram. A free body is a piece of a structure we take out and make free to see the forces acting internally in the structure. Note that we don't want a free body to go spinning off into space and lose it. The structure which contains the free body was in equilibrium so the portion of the structure we have removed should also be. The addition of the support reactions and external loads on the free body at the locations of cut members maintains static equilibrium. (And if it is in static equilibrium our equations of static equilibrium apply to the free body). A free body diagram can be used to find forces on the structure, like reactions, or can be used to find internal forces.

The principles of boundary conditions and free body diagrams are easily shown in examples:

1. Find the reactions of a beam supported by a pin and a roller on opposite ends, 20 feet in length, which contains a load at the center point of 60 kips.

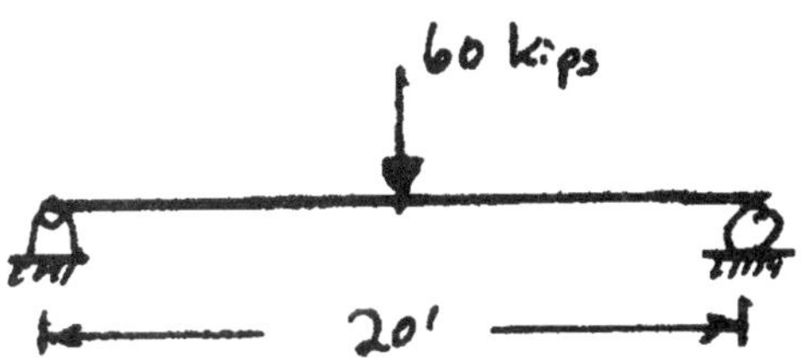

Figure 6.4 Example Problem 1

From Example problem 1 a free body diagram showing forces can be shown

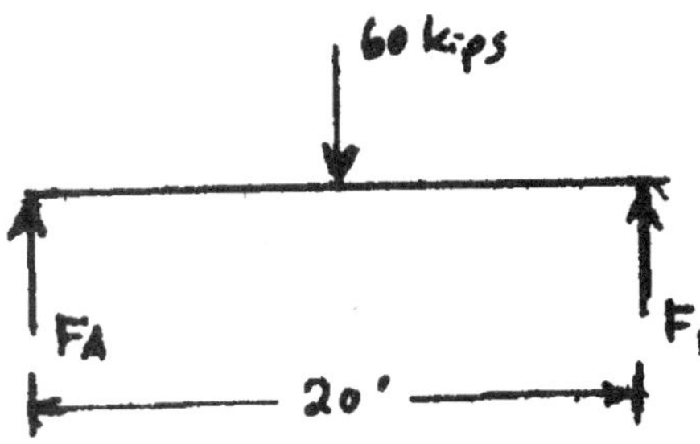

Figure 6.5 Freebody Diagram - Example 1

To solve for the reactions at A and B use our three equations of equilibrium. First, sum moments about the pin at A to solve for the vertical reaction at B.

Sum of moments about A = 0
Sum $M_A = 0$

Since the pin at A cannot resist moments we know that the moment at the pin is 0. Note that the forces point in different directions around the pin, and thus we have a different sign. The sign convention is discussed in the next section. We also know that since the horizontal forces at reactions A and B act directly through point A they cannot produce a moment about A (i.e. M = F x d since d = 0 then M = 0). This leaves us with only the vertical force at B as an unknown in our equation.

Sum $M_A = 0$
$(-60 \text{ kips}) \times (10 \text{ ft}) + (F_{yB}) \times (20 \text{ ft}) = 0$
$-600 \text{ kip ft} + (F_{yB}) \times (20 \text{ ft}) = 0$
$(F_{yB}) \times (20 \text{ ft}) = 600 \text{ kip ft}$
$F_{yB} = (600 \text{ kip ft}) / (20 \text{ ft})$

$F_{yB} = 30 \text{ kips}$

Now to solve for the vertical reaction at A we can use another of our equations of equilibrium.

Sum of forces in y direction = 0
Sum $F_y = 0$
$F_{yA} - 60 \text{ kips} + F_{yB} = 0$
$F_{yA} - 60 \text{ kips} + 30 \text{ kips} = 0$
$F_{yA} - 30 \text{ kips} = 0$
$F_{yA} = 30 \text{ kips}$

By observation we know that the horizontal reaction at A in this example must be zero, but let us use our third equation of equilibrium to show this.

Sum of forces in x direction = 0
Sum $F_x = 0$
$F_{xA} + F_{xB} = 0$

Since the roller at B cannot resist a horizontal force we know $F_{xB} = 0$. Thus our equation becomes:

$F_{xA} + 0 = 0$
$F_{xA} = 0$

Free bodies created within structures are also easily demonstrated with examples:

2. Find the force in a bar suspended from a pinned boundary condition created by a force of 60 kips.

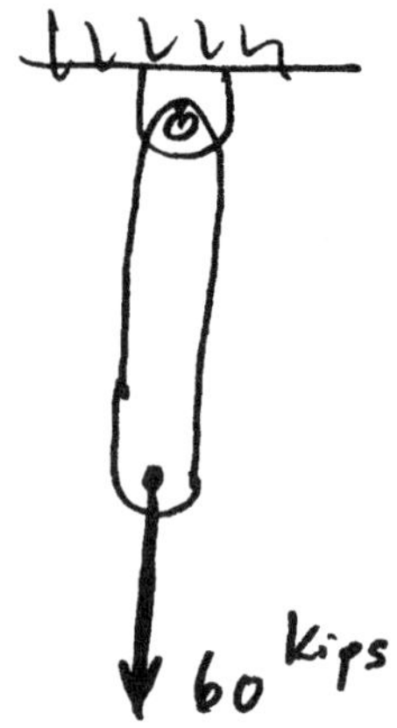

Figure 6.6 Example 2

This one is easy to show the idea:

Figure 6.7 Free body example 2

From the free body diagram it is obvious that the internal force is 60 kips but let us demonstrate this using one of our equations of equilibrium.

Sum of forces in y direction = 0
Sum $F_y = 0$
$F_y - 60 \text{ kips} = 0$
$F_y = 60 \text{ kips}$

Note that since the force acts to stretch the member the member is in tension.

6.6 Determinant Trusses - Methods of Pins and Sections

We can analyze trusses with two techniques using free bodies and the principles of static equilibrium. One is called the method of pins the other is the method of sections.

6.6.1 The method of pins

The method of pins looks at the forces that act on an individual pin. Since a pin connection cannot resist a moment only the other two equations of equilibrium will be helpful (Sum $F_x = 0$, Sum $F_y = 0$). The analysis is begun from a location where there is a known force and we can solve 1 unknown at a time, up to two unknowns using the equations of equilibrium.

Let's look at an example:

Find the force in all the members created by a 1000 lb. downward force applied at C on the simple truss shown below:

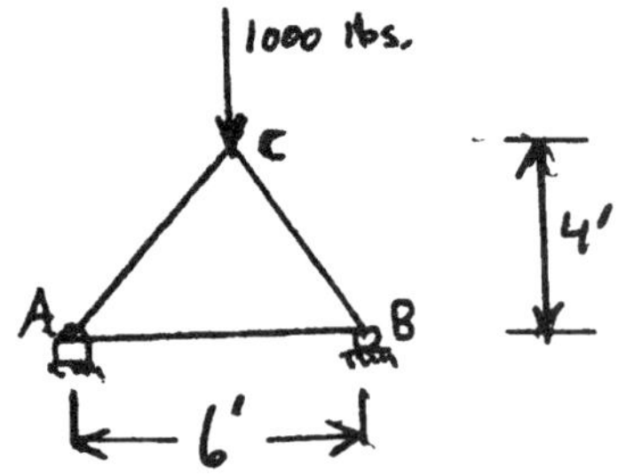

Figure 6.8 Example 3

This one is a little tougher than the previous example, but if we break it down into small steps and solve one step at a time it becomes more manageable.

First we find the support reactions at A and B:

To find the y direction force at B we can sum the moments about A. (Note that by summing moments about A we do not have to be concerned with the unknown x direction forces at A and B or the unknown y direction force at A since all of these forces act in direct line through point A their moment arm is 0 thus the moment they produce about A is 0).

Sum of moments about A = 0
Sum $M_A = 0$
$(1000 \text{ lbs}) \times (3 \text{ ft}) - (F_{y\ Reaction\ B}) \times (6 \text{ ft}) = 0$
$3000 \text{ ft lbs} - (F_{y\ Reaction\ B}) \times (6 \text{ ft}) = 0$
$F_{y\ Reaction\ B} = 3000 \text{ ft lbs} / 6 \text{ ft}$
$F_{y\ Reaction\ B} = 500 \text{ lbs}$

Now using the sum of the forces in the y direction we can solve for the y reaction at A:
Sum of forces in y direction = 0
Sum $F_y = 0$
$F_{yA} - 1000 \text{ lbs} + F_{yB} = 0$
$F_{yA} - 1000 \text{ lbs} + 500 \text{ lbs} = 0$
$F_{yA} - 500 \text{ lbs} = 0$
$F_{yA} = 500 \text{ lbs}$

By inspection we can determine that the x direction forces at both A and B must be 0 (see discussion in example 1).

Now using the method of pins we can solve for the member forces. Pick a pin with one known force and two or less unknowns. Let's start with pin A. We do not know the forces in member AC or AB but we do know the direction those forces act because of the geometry of the truss. We can thus solve for these member forces one at a time using the equations of equilibrium. Only member force AC has a y force component so we will begin with this calculation. We need the force vector in the y direction which is $4/5\ F_{A\text{-}C}$ (recognizing that we have a 3-4-5 triangle).

Sum of forces in y direction = 0
$F_y = 0$
F_{yA} + (y component of force AC) = 0
$500 \text{ lbs} + (4/5)\ F_{AC} = 0$
$(4/5)\ F_{AC} = -500 \text{ lbs}$
$F_{AC} = (5/4) \times (-500 \text{ lbs})$
$F_{AC} = -2500 \text{ lbs}/4$
$F_{AC} = -625 \text{ lbs}$

Note that we have come up with a negative result. This indicates that the actual force in member AC acts in the opposite direction as assumed. The sign convention used here is that tension forces within the member are considered positive. All unknown forces are assumed to be positive (tensions), for convenience, until their actual value is known.

The free body diagram can be a little confusing at first as to sign convention. This is because there is an arrow pointing away from the pin which at first appears to put the member in compression. Two ways to think about the sign convention is to picture the cut through the member; or to think of tension in the pin is tension in the member, see diagram in Figure 6.9.

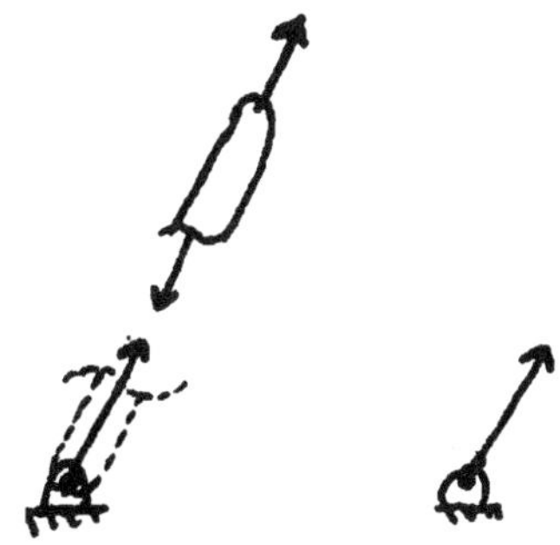

Figure 6.9 Sign convention (tension) in method of pins

When calculating moments, there is also a sign How do we know which is positive and which is negative? Engineers made a rule for everyone to follow called the Right Hand Sign Rule. Make a fist with your right hand with your thumb pointing up. If loads rotate around a point like your fingers, the moment will be called positive (up). If loads go in the opposite direction (down), moments are said to be negative.

Continuing with the solution for the member forces in example 3:

Again looking at Pin A we can now solve for the force in member AB.

Sum of forces in x direction = 0
Sum $F_x = 0$
F_{xA} + (x component of force AC) + $F_{AB} = 0$
$0 + (3/5)\ F_{AC} + F_{AB} = 0$
$0 + (3/5) \times (-625 \text{ lbs}) + F_{AB} = 0$
$0 - 375 \text{ lbs} + F_{AB} = 0$
$F_{AB} = 375 \text{ lbs}$

(Note that F_{AB} is positive thus the member is in tension)

The same process can be continued from pin C or pin B to find the force in member BC (F_{BC}). However, note that the structure is symmetrical in geometry and loading. Symmetrical structures that are symmetrically loaded have symmetrical internal forces. Thus, the force in member BC is the same as the force in member AC, 625 lbs compression.

Larger trusses can be analyzed using the same technique. Take for example the truss below. What is the force in each member?

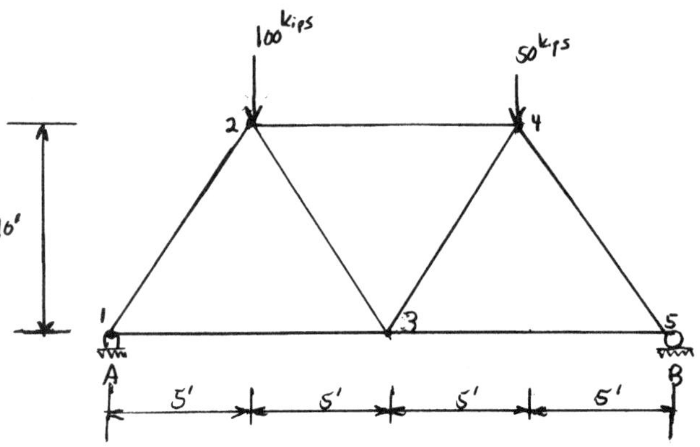

Figure 6.10 - Example Problem for Truss Analysis

Figure 6.11 which follows shows a typical worksheet to find forces. The process is the same as in the last example.

6.6.2 The method of sections

If we need to know the force in a single member of a truss, we can quickly find it by sawing out a section through that member to create a free body and use our equations of equilibrium to find the forces. Find the force in member 3-4 of the previous problem.

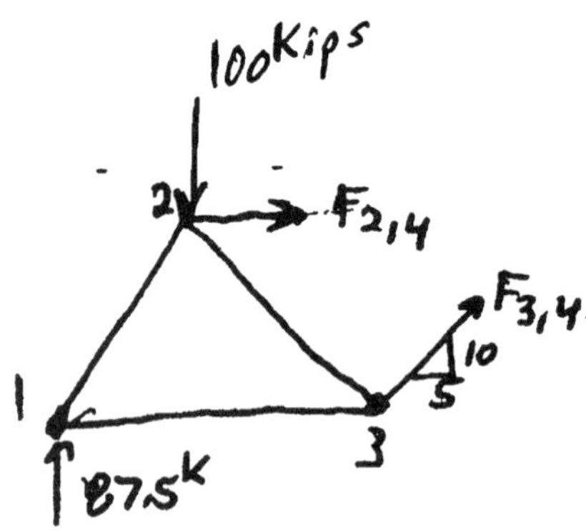

Figure 6.12 Free body method of sections

At point 3
Sum $F_y = 0$

$10/11.2\ F_{3,4} - 100 + 87.5 = 0$
$F_{3,4} = 13.98K$ Tension

At Point 1
Sum M = 0

$-100(5) + (10/11.2)13.98(10) - F_{2,4}(10) = 0$
$F_{2,4} = -37.5K = 37.5$ K Compression

6.7 Indeterminant Trusses - Springs and Finite Element Method

Indeterminant trusses are created by structures which can not be solved by the principles of static equilibrium already described. That is, there are too many unknowns on any particular freebody to solve using the three equations of static equilibrium. Note that structures will still be in static equilibrium, but determining the forces on any given member may not be possible. This problem can be resolved with the use of a systematic approach that relies on the solution of multiple simultaneous equations to find a solution to all structural variables at once. The concepts were developed as one of the earliest findings in structural analysis and were advanced with the advent of modern computers.

6.7.1 The Spring Law or The Stiffness Law

Robert Hooke is credited with being one of the first researchers to study springs. This led to the study of the stiffness of structures in approximately 1660. His work was done to understand the response of metal rods for use in ship building. The work seems simple to us today, but 300 years ago it was a breakthrough. With the advent of modern computers, this concept became one of the most important and fundamental principles in structural engineering.

The spring is a familiar structural member. We are familiar with spring elements such as are found on screen doors or other household items. If we hang a spring on a hook and add a weight from the bottom, the obvious, expected result is that the load will cause the spring to stretch or deflect. Similiarily, if we replace the spring with a heavy steel bar, the same thing will happen- the steel bar will stretch.

1. Find reactions for structure:

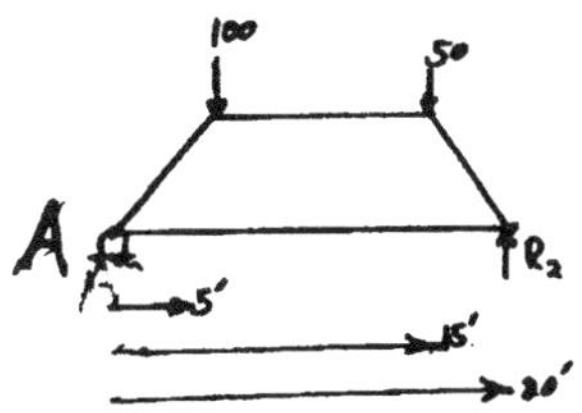

Sum $M_{R\text{-}A} = 0$

$-100(5)-50(15)+R_2(20)=0$

$R_2 = 62.5$ Kips

Sum $F = 0$

$-100-50+62.5+R_1 = 0$

$R_1 = 87.5$ K

2. Cut a freebody through pin 1

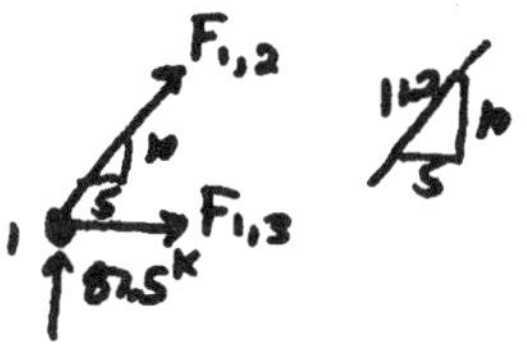

Sum $F_y = 0$

$10/11.2\ F_{1,2} + 87.5 = 0$

$F_{1,2} = -98$ K = 98 K Compression

Sum $F_x = 0$

$F_{1,3} - 98(5/11.2) = 0$

$F_{1,3} = 43.75$ K = 43.75 K Tension

3. For pin 2

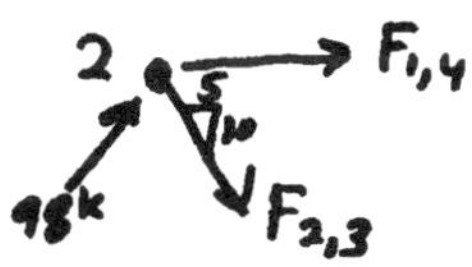

Sum $F_y = 0$

$-100 + 98(10/11.2) - (10/11.2)\ F_{2,3} = 0$

$F_{2,3} = -13.98$ K = 13.98 K Compression

Sum $F_x = 0$

$+98(5/11.2) - 13.98(5/11.2) + F_{2,4} = 0$

$F_{2,4} = -37.52$ K = 37.52 K Compression

4. For pin 3

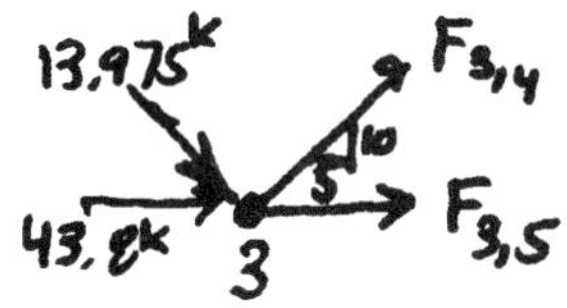

Sum $F_y = 0$

$-13.975(10/11.2)+(10/11.2)F_{3,4} = 0$

$F_{3,4} = 13.975$K = 13.975 K Tension

Sum $F_x = 0$

$F_{3,5} + 13.975(5/11.2) - 43.8 + (5/11.2)(13.975) = 0$

$F_{3,5} = 31.3$ K = 31.3 K Tension

5. Finish with pin 5

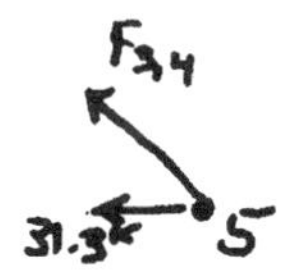

Sum $F_y = 0$

$62.5+(10/11.2)\ F_{5,4} = 0$

$F_{5,4} = -69.875$ K = 69.875 K Compression

Figure 6.11 - Solution of a truss using the Method of Pins

Of course, the spring will stretch more than the steel bar because the steel bar is stiffer than the spring.

This relationship between load, P, and deflection, y, is common between all structures and therefore, is noted as a structural law, the Spring Law, or Law of Stiffness.

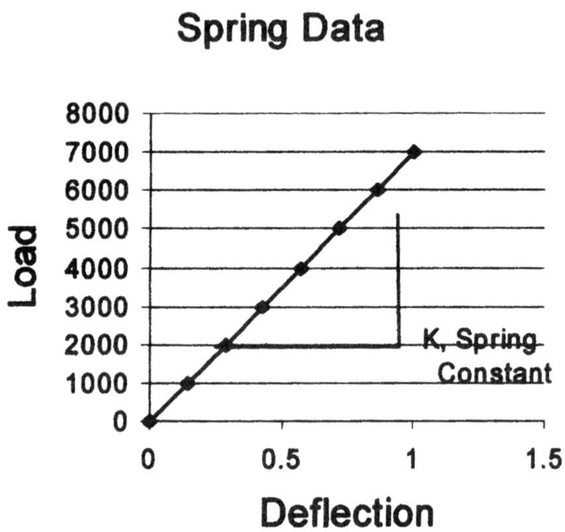

Figure 6.13 - Graphical relationship of Spring Stiffness.

Stiffness Law : All structures when subjected to load will encounter displacement. The displacement/ load relationship may generally be related through a constant, the stiffness constant, such that:

K y = P

Where:
K = the stiffness constant (in one dimensional problems), or matrix of constants (in multi-dimensional problems)

y = displacement of the structure

P = load

This relationship may also be seen graphically as shown in Figure 6.13.

The Spring Constant is expressed in the units of pounds per inch (Lbs/in). For example, if a spring has a constant, K=5, it takes 5 pounds to deflect the spring 1 inch.

6.7.2 Multiple spring systems

If more than one spring were to be used, the deflection would be carried or reduced by the multiple springs. First consider two springs of equal spring constant hung side by side to carry a load. Now the stiffness resisting the load is (K+K)*y = P. Or, the deflection y will be P/(2K), or one half as much. If the two springs were hung at an angle to the load, say 45°, then the force vector of the springs must be taken into account, or (1/1.41 K+ 1/1.41 K) y = P. And thus y = P/(1.414K) .

Truss members act as springs due to the force being carried along the axis of the members. Thus, the stiffness of trusses, derived in the next chapter, can be added together and all unknown displacements solved simultaneously.

6.7.3 Introduction of the Finite Element Method

The Finite Element Method is a mathematical technique which uses the Spring Law for structures by dividing the structure into individual members and adds the stiffness contribution from each member to solve for the behaviour of the entire structure. For trusses this is easy since each member is treated separately as a spring.

The Finite Element Method, or sometimes called the Stiffness Method, can be readily programmed for solution on the computer. Such a computer program is contained at the end of this section. The process to solve a truss using this method is similar to the method of pins. Each point where members intersect is defined as a node and the x,y coordinates of each node are used as input. Elements are named using the node at each end of the member. Loads and boundary conditions are defined and also used as input. Boundary conditions in computer analysis are very important to the solution. The structure must have boundary conditions that will allow the structure to be statically stable for all loading conditions, irregardless of what loading conditions are used for input.

To do this at least one pin and one roller are generally used for trusses in a simple span. (More information on structural stability is contained in Chapter 8).

6.8 Computer Program TF2DFE

A program, TF2DFE, was written to follow the concepts described above. BASIC language is used as it is readily available on most computers. The program uses a simultaneous equation solver, (see some fun mathematics in the Workbook Appendix), which is limited to a total of 80 unknowns (displacment directions). Fifty nodes and 100 elements may be used.

This program is written to use boundary conditions and load in the x,y direction, called the global direction, x is referred to the program as direction, 1, with y being direction, 2. For our example problem, the diagram to produce input to the program would appear as shown in Figure 6.14. The example problem is used to demonstrate the use of the program. A line by line description of input follows.

With standard input prepared (see next section), any problem can be solved by using the DATA statements with the computer program. Note that all units must be the same within the program. Loads, if input in pounds, will result in element forces being printed which are in pounds. Coordinates can be multiplied by a length factor for convenience, for example, length input in feet but material properties entered in inches and displacements in inches. For first time users it is probably easier to enter coordinates in feet and use the factor as twelve, or in metric enter in meters and use the factor as one. Loads would be entered in pounds or kilonewtons.

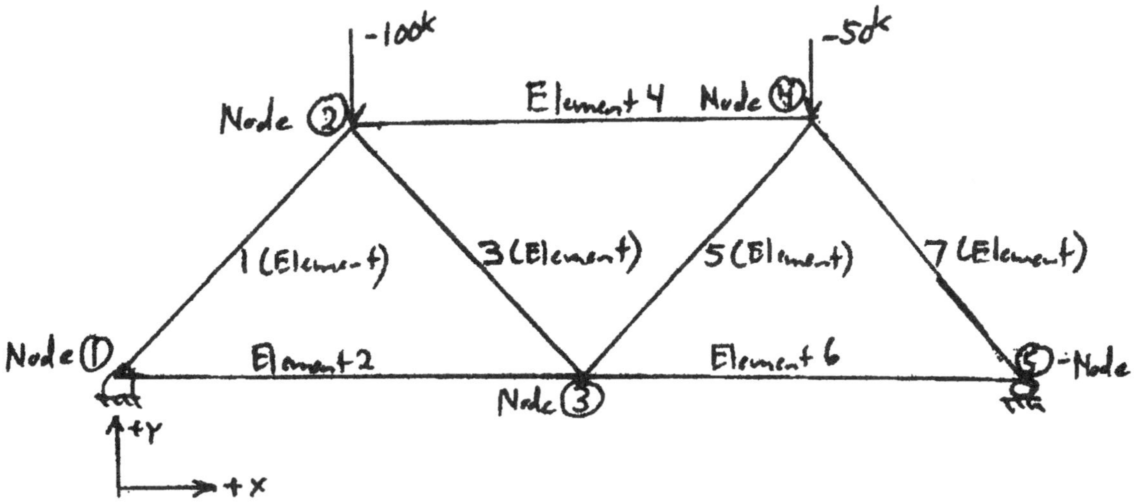

Figure 6.14 Example Problem labeled for Finite Element Solution

Notes:

Boundary condition at node 1 (Pin) - (Fixed in direction 1 (x) and 2 (y))
Boundary condition at node 2 (roller).Fixed in direction 2 (y)
Loads for this example are in direction 2 (y) - Note value of load is negative (in - y direction)
Load values will be positive or negative
Load direction will always be positive.

Program TF2DFE Input instructions

The full data set for the problem in figure 6.14 is shown below:

```
19999 REM DATA STATEMENTS FOR PROBLEM
20000 REM
20010 REM TITLE FOR PROBLEM
20020 DATA "  WARREN TRUSS EXAMPLE"
20030 REM NUMBER OF NODES, LENGTH FACTOR
20040 DATA 5,12
20050 REM X,Y COORDINATES (UNITS = FEET)
20060 DATA 0,0
20070 DATA 5,10
20080 DATA 10,0
20090 DATA 15,10
20100 DATA 20,0
20110 REM MATERIALS- NUMBER
20120 DATA 1
20130 REM MATERIALS VALUES: AREA,MODULUS (UNITS=INCHES)
20140 DATA 10,36000000
20150 REM ELEMENTS- NUMBER
20160 DATA 7
20170 REM NODE1, NODE2, MATERIAL TYPE  OF ELEMENT
20180 DATA 1,2,1
20190 DATA 1,3,1
20200 DATA 2,3,1
20210 DATA 2,4,1
20220 DATA 3,4,1
20230 DATA 3,5,1
20240 DATA 4,5,1
20250 REM BOUNDARY CONDITIONS - NUMBER
20260 DATA 3
20270 REM BC'S NODE NUMBER, DIRECTION (X=1, 2=Y)
20280 DATA 1,1
20290 DATA 1,2
20300 DATA 5,2
20310 REM   LOADS - NUMBER
20320 DATA 2
20330 REM NODE, DIRECTION,   LOAD VALUE
20340 DATA 2,2,-100000
20350 DATA 4,2,-50000
```

Input data for Problems

The description of the input required for use of the program is shown below.

Line 1. Title for this problem:
DATA "WARREN TRUSS EXAMPLE"

Line 2. Number of nodes, Length factor

The length factor is used to multiply the input coordinates for use within the program. It is supplied for the convenience of the user. The code assumes that all units are the same. A user can have dimensions in the data file of feet, and use the factor as 12 to use inches in the code and the output. (for example the material properties, (see next chapter) are in inches and displacements are desired to be in inches, but input is more convenient in feet, use 12) :
DATA 3,12

Line 3. x,y Coordinates for each node:
DATA 0,0
DATA 0,10
.
List all data

Line 4. Number of different types of material (see next chapter for explanation of material properties):
DATA 1

Line 5. Area, Modulus of Elasticity(see next chapter):
DATA 10,36000000

Line 6. Number of elements in structure
DATA 5

Line 7. Element end node numbers, e.g. 1 and 2, and material number from above:
Note end node numbers need to be smallest, largest
DATA 1,2,1
DATA 1,3,1
.
For number of elements

Line 8. Number of boundary conditions
see notes, node 1 has 2 boundary conditions, node 2 has 1 boundary condition:
DATA 3

Line 9. Node number, Direction in which displacement can not occur (x direction =1, y direction =2):
DATA 1,1
DATA 1,2
DATA 2,1
.
For the number of boundary conditions

Line 10. Number of applied loads:
DATA 4

Line 11. Node number, Global Direction, Value of load (+ in plus coordinate direction), example:
DATA 1,2,-100
DATA 2,2,100
For the number of loads

Truss Fun

Program TF2DFE Selected Output

Displacements

Node	X	Y
1	0	0
2	1.98E-02	-5.06E-02
3	1.45E-02	-4.74E-02
4	7.29E-03	-3.79E-02
5	0.025	0

Displacements are in units of input by node number specified by user

X	is displacement x (horizontal) direction
Y	is displacement in y (vertical) direction
	Negative displacements are down (y) or to the left (x) traditionally

Element Summary

ELE#	N1	N2	A	E	L	IL	F	STR
1	1	2	10	36000000	134.2	11.2	-97828	-9782.8
2	1	3	10	36000000	120	10	43750	4375
3	2	3	10	36000000	134.2	11.2	-13975	-1397.4
4	2	4	10	36000000	120	10	-37500	-3750.0
5	3	4	10	36000000	13.2	11.2	13975	1397.5
6	3	5	10	36000000	120	10	31250	3125
7	4	5	10	36000000	134.2	11.2	-69877	-6987.7

For the element summary table the following data is contained in the columns:

Column Heading	Description of Output found
ELE#	Element number from user input
N1	Beginning node from user input
N2	Ending node from user input
A	Area for element from user input
E	Young's Modulus for element from user input
L	Calculated length of element in units used in all calculations (Input units* Length Factor)
IL	Length of element in units input by user
F	Calculated force in member (element) in consistent units with input loads and properties
STR	Calculated stress in member (element) in consistent units with input loads and properties

The Program Listing: TF2DFE
(Can be typed, or see order blank in back for CD)

```
1 REM  Finite Element Program
2 REM For Planar Trusses,
3 REM to be used only for study
4 REM not guaranteed for analysis
5 REM or design.  Uses global
6 REM  loads, square assembly,
7 REM in core solver,w/
8 REM compressed banded matrix.
9 REM David W. Harris, PhD,PE
100 REM PROBLEM SIZE - USER CHANGEABLE
110 REM TITLE
120 DIM T$(20)
130 REM G NEEDS TO BE SQUARE FOR ASSEMBLY
140 REM DOUBLE PRECISION RECOMMENDED
150 REM FOR ACCURACY IN COMPUTATION
160 DEFDBL A-Z
170 DIM G(80, 80), Q(80)
180 REM NODES
190 DIM X(50), Y(50)
200 REM MATERIALS
210 DIM A(10), E(10)
220 REM ELEMENTS
230 DIM E1(100), E2(100), M(100)
240 REM BOUNDARY CONDITIONS
250 DIM BC(20)
260 REM  LOADS
270 DIM P(100), B(100)
280 REM PROGRAM VARIABLES
290 REM ELEMENT STIFFNESS
300 DIM EE(4, 4), EF(4), SE(4), II(4)
310 REM MAXIMUM NUMBER VARIABLES
320 REM MAXIMUM NUMBER OF LINES ON SCREEN
330 N6 = 15
340 REM TOTAL DEGREES OF FREEDOM = 80
350 N50 = 80
360 REM MAXIMUM NODES = 40
370 N10 = 40
380 REM MAXIMUM MATERIALS = 10
390 N20 = 10
400 REM MAXIMUM ELEMENTS = 100
410 N30 = 100
420 REM MAXIMUM BOUNDARY CONDITIONS = 20
430 N40 = 20
440 REM MAXIMUM  LOADS = 100
450 N59 = 100
460 REM FOR USE WITHOUT DOUBLE PRECISION
470 REM RULE OF THUMB ACCURACY
480 AC = .00001
1000 REM MAIN PROGRAM
1010 REM READ INPUT DATA
1020 GOSUB 2000
1030 REM ASSEMBLE STIFFNESS MATRIX
1040 GOSUB 3200
1050 REM ASSEMBLE  LOADS
1060 GOSUB 4000
1070 REM APPLY BOUNDARY CONDITIONS
1080 GOSUB 5000
1090 REM SHIFT MATRIX TO COMPRESSED FORM
1100 GOSUB 5500
1110 REM SOLVE SIMULTANEOUS EQUATIONS
1120 GOSUB 6000
1130 REM PRINT DISPLACEMENTS
1140 GOSUB 7000
1150 REM PRINT ELEMENT FORCES AND STRESSES
1160 GOSUB 8000
1170 PRINT
1180 PRINT "  END OF SOLUTION"
1190 PRINT
1999 END
2000 REM INPUT SUBROUTINE
2010 REM TITLE
2020 PRINT
2030 PRINT
2040 PRINT
2050 PRINT "******** NEW PROBLEM **********"
2060 PRINT
2070 READ T$
2080 PRINT T$
2090 REM NODES AND FACTOR TO CONVERT LENGTH
2100 READ n1, LF
2110 IF n1 < N10 GOTO 2140
2120 PRINT "NUMBER OF NODES EXCEEDS CODE MAXIMUM"
2130 END
2140 PRINT " N          X          Y   FOR", n1, "NODES"
2150 N99 = 1
2160 FOR I = 1 TO n1
2170 READ X(I), Y(I)
2180 PRINT I, X(I), Y(I)
2190 N99 = N99 + 1
2200 IF N99 < N6 GOTO 2240
2210 INPUT ; "TOUCH ANY KEY TO CONTINUE", Z$
2220 N99 = 1
2230 PRINT
2240 NEXT I
2250 INPUT ; "TOUCH ANY KEY TO CONTINUE", Z$
2260 PRINT
2270 PRINT "N          X          Y CONVERTED FOR"; n1; "
NODES"
2280 N99 = 1
2290 FOR I = 1 TO n1
2300 X(I) = X(I) * LF
2310 Y(I) = Y(I) * LF
2320 PRINT I, X(I), Y(I)
2330 N99 = N99 + 1
2340 IF N99 < N6 GOTO 2380
2350 INPUT ; "TOUCH ANY KEY TO CONTINUE", Z$
2360 PRINT
2370 N99 = 1
2380 NEXT I
2390 INPUT ; "TOUCH ANY KEY TO CONTINUE", Z$
2400 REM MATERIALS
2410 READ N2
2420 IF N2 < N20 GOTO 2460
2430 PRINT
2440 PRINT "NUMBER OF MATERIALS EXCEEDS CODE MAXI-
MUM"
2450 END
2460 PRINT
2470 PRINT
2480 PRINT "IM  AREA      MODULUS    FOR", N2, "MATERIALS"
2490 FOR I = 1 TO N2
2500 READ A(I), E(I)
2510 PRINT I; A(I), E(I)
2520 NEXT I
2530 INPUT ; "TOUCH ANY KEY TO CONTINUE", Z$
2540 REM ELEMENTS
2550 READ N3
```

```
2560 IF N3 < N30 GOTO 2600
2570 PRINT
2580 PRINT "NUMBER OF ELEMENTS EXCEEDS CODE MAXI-
MUM"
2590 END
2600 PRINT
2610 PRINT
2620 PRINT " IE NODE1     NODE2     MATERIAL FOR", N3, "
ELEMENTS"
2630 N99 = 1
2640 FOR I = 1 TO N3
2650 READ E1(I), E2(I), M(I)
2660 PRINT I; E1(I), E2(I), M(I)
2670 N99 = N99 + 1
2680 IF N99 < N6 GOTO 2720
2690 INPUT ; "TOUCH ANY KEY TO CONTINUE"; Z$
2700 PRINT
2710 N99 = 1
2720 NEXT I
2730 INPUT ; "TOUCH ANY KEY TO CONTINUE", Z$
2740 REM BOUNDARY CONDITIONS
2750 READ N4
2760 IF N4 < N40 THEN 2800
2770 PRINT
2780 PRINT "NUMBER OF BOUNDARY CONDITIONS EXCEEDS
CODE MAXIMUM"
2790 END
2800 PRINT
2810 PRINT
2820 PRINT " NODE     DIRECTION     EQ # FOR", N4, "BOUND-
ARY CONDITIONS "
2830 FOR I = 1 TO N4
2840 READ B, D
2850 BC(I) = ((B - 1) * 2) + D
2860 PRINT B, D, BC(I)
2870 NEXT I
2880 INPUT ; "TOUCH ANY KEY TO CONTINUE", Z$
2890 REM     LOADS
2900 READ N5
2910 IF N5 < N59 THEN 2960
2920 REM
2930 PRINT
2940 PRINT "NUMBER OF     LOADS EXCEEDS CODE
MAXIMUM"
2950 END
2960 PRINT
2970 PRINT
2980 PRINT " NODE     DIRECTION          LOAD     EQ
NUMBER FOR", N5; "     LOADS"
2990 N99 = 1
3000 FOR I = 1 TO N5
3010 READ B, D, P99
3020 P(I) = P99
3030 B(I) = ((B - 1) * 2) + D
3040 PRINT B, D, P(I), B(I)
3050 N99 = N99 + 1
3060 IF N99 < N6 GOTO 3100
3070 INPUT ; "TOUCH ANY KEY TO CONTINUE", Z$
3080 PRINT
3090 N99 = 1
3100 NEXT I
3110 INPUT ; "TOUCH ANY KEY TO CONTINUE", Z$
3199 RETURN
3200 REM ASSEMBLE MATRICES
3202 REM  STIFFNESS
3204 REM INITIALIZE MATRICES
3206 FOR I = 1 TO N50
3208 FOR J = 1 TO N50
3210 G(I, J) = 0
3212 Q(I) = 0
3214 NEXT J
3216 NEXT I
3218 REM ASSEMBLE STIFFNESS ELEMENT BY ELEMENT
3220 FOR I = 1 TO N3
3222 GOSUB 3228
3224 GOTO 3500
3226 REM CREATE ELEMENT STIFFNESS
3228 I1 = E1(I)
3230 I2 = E2(I)
3232 DX = X(I2) - X(I1)
3234 DY = Y(I2) - Y(I1)
3236 XL = SQR(DX * DX + DY * DY)
3238 IF DX = 0 GOTO 3244
3240 THC = DY / DX
3242 GOTO 3248
3244 TH = 90 * 3.1416 / 180 * ABS(DY) / DY
3246 GOTO 3250
3248 TH = ATN(THC)
3250 C = DX / XL
3260 S = DY / XL
3270 CC = C * C
3280 CS = C * S
3290 SS = S * S
3300 IM = M(I)
3310 YM = E(IM)
3320 AR = A(IM)
3330 EE(1, 1) = YM * AR / XL * CC
3340 EE(1, 2) = YM * AR / XL * CS
3350 EE(1, 3) = -YM * AR / XL * CC
3360 EE(1, 4) = -YM * AR / XL * CS
3370 EE(2, 1) = EE(1, 2)
3380 EE(3, 1) = EE(1, 3)
3390 EE(4, 1) = EE(1, 4)
3400 EE(2, 2) = YM * AR / XL * SS
3410 EE(2, 3) = -YM * AR / XL * CS
3420 EE(2, 4) = -YM * AR / XL * SS
3430 EE(3, 2) = EE(2, 3)
3440 EE(4, 2) = EE(2, 4)
3450 EE(3, 3) = EE(1, 1)
3460 EE(3, 4) = YM * AR / XL * CS
3470 EE(4, 3) = EE(3, 4)
3480 EE(4, 4) = EE(2, 2)
3499 RETURN
3500 REM PLACE INTO GLOBAL STIFFNESS
3510 REM ASSEMBLE IN SQUARE MATRIX
3520 REM  MATRIX IS BANDED
3530 IG = (I1 - 1) * 2
3540 JG = (I2 - 1) * 2
3550 FOR K = 1 TO 2
3560 FOR L = 1 TO 2
3570 KP = K + 2
3580 LP = L + 2
3590 G(IG + K, IG + L) = G(IG + K, IG + L) + EE(K, L)
3600 G(IG + K, JG + L) = G(IG + K, JG + L) + EE(K, LP)
3610 G(JG + K, IG + L) = G(JG + L, IG + L) + EE(KP, L)
3620 G(JG + K, JG + L) = G(JG + K, JG + L) + EE(KP, LP)
3630 NEXT L
3640 NEXT K
3650 NEXT I
3999 RETURN
4000 REM ASSEMBLE     LOADS
```

```
4010 FOR I = 1 TO N5
4020 I1 = B(I)
4030 Q(I1) = Q(I1) + P(I)
4040 NEXT I
4050 RETURN
5000 REM APPLY BOUNDARY CONDITIONS TO STIFFNESS
5010 KK = n1 * 2
5020 FOR K = 1 TO N4
5030 FOR I = 1 TO KK
5040 BV = 0
5050 J = BC(K)
5060 Q(I) = Q(I) - G(I, J) * BV
5070 NEXT I
5080 NEXT K
5090 FOR I = 1 TO N4
5100 N = BC(I)
5110 FOR J = 1 TO KK
5120 G(N, J) = 0
5130 NEXT J
5140 FOR J = 1 TO KK
5150 G(J, N) = 0
5160 NEXT J
5170 G(N, N) = 10000
5180 NEXT I
5499 RETURN
5500 REM COMPRESS MATRIX FOR SOLVER
5510 REM SHIFT DIAGONAL TERM TO COL. 1
5520 REM BANDWIDTH
5530 MM = 0
5540 FOR IE = 1 TO N3
5550 ID1 = ABS(E1(IE) - E2(IE)) * 2
5560 IF ID1 < MM GOTO 5580
5570 MM = ID1
5580 NEXT IE
5590 MM = MM + 2
5600 REM SHIFT MATRIX
5610 NN = n1 * 2
5620 FOR I = 1 TO NN
5630 K = I + MM
5640 IF K < NN THEN 5660
5650 K = NN
5660 FOR J = I TO K
5670 KK = J - I + 1
5680 G(I, KK) = G(I, J)
5690 NEXT J
5700 NEXT I
5710 IB = MM + 1
5720 FOR I = 1 TO IB
5730 II = NN - I + 1
5740 FOR J = I TO IB
5750 JJ = J + 1
5760 G(II, JJ) = 0
5770 NEXT J
5780 NEXT I
5799 RETURN
6000 REM SIMULTANEOUS EQUATION SOLVER
6010 REM
6080 NN = n1 * 2
6090 FOR N = 1 TO NN
6100 IF ABS(G(N, 1)) < .00001 GOTO 6450
6110 FOR L = 2 TO MM
6120 IF G(N, L) <> 0 GOTO 6150
6130 C = 0
6140 GOTO 6260
6150 C = G(N, L) / G(N, 1)
6160 IF ABS(C) < .00001 GOTO 6180
6170 GOTO 6190
6180 C = 0
6190 I = N + L - 1
6200 IF (NN - I) < 0 GOTO 6260
6202 IF (NN - I) = 0 GOTO 6216
6204 IF (NN - I) > 0 GOTO 6216
6216 J = 0
6220 FOR K = L TO MM
6222 J = J + 1
6224 G(I, J) = G(I, J) - C * G(N, K)
6250 NEXT K
6252 IF ABS(Q(N)) < .00001 GOTO 6254
6253 GOTO 6255
6254 Q(N) = 0
6255 Q(I) = Q(I) - C * Q(N)
6258 G(N, L) = C
6260 NEXT L
6270 Q(N) = Q(N) / G(N, 1)
6280 NEXT N
6290 REM  BACK SUBSTITUTION
6292 N = NN
6300 N = N - 1
6302 IF N < 0 GOTO 6500
6304 IF N = 0 GOTO 6500
6306 IF N > 0 GOTO 6350
6350 FOR K = 2 TO MM
6360 L = N + K - 1
6362 IF (NN - L) < 0 GOTO 6440
6364 IF (NN - L) = 0 GOTO 6401
6366 IF (NN - L) > 0 GOTO 6401
6401 IF ABS(Q(L)) < .00001 GOTO 6403
6402 GOTO 6404
6403 Q(L) = 0
6404 IF ABS(G(N, K)) < .00001 GOTO 6406
6405 GOTO 6407
6406 G(N, K) = 0
6407 Q(N) = Q(N) - Q(L) * G(N, K)
6440 NEXT K
6442 GOTO 6300
6450 PRINT "BAD DIAGONAL IN EQ.", N
6452 STOP
6500 RETURN
7000 REM PRINT DISPLACEMENTS
7010 TA = 10000
7020 IN = 1
7030 PRINT
7040 PRINT "  SOLUTION"
7050 PRINT
7060 PRINT "NODE        X-DISP        Y-DISP"
7070 PRINT
7080 IC = 5
7090 FOR I = 1 TO NN STEP 2
7100 Q! = Q(I)
7110 QP! = Q(I + 1)
7120 PRINT IN, Q!, QP!
7130 IC = IC + 1
7140 IF Q(I) = 0 GOTO 7170
7150 IF ABS(Q(I)) > TA GOTO 7170
7160 TA = Q(I)
7170 IF Q(I + 1) = 0 GOTO 7200
7180 IF ABS(Q(I + 1)) > TA GOTO 7200
7190 TA = Q(I + 1)
7200 IF IC < N6 GOTO 7260
7210 INPUT ; " TOUCH ANY KEY TO CONTINUE", Z$
```

```
7220 PRINT
7230 IC = 3
7240 PRINT T$
7250 PRINT "NODE        X-DISP          Y-DISP"
7260 IN = IN + 1
7270 NEXT I
7280 IF ABS(TA) > AC GOTO 7999
7290 PRINT "  **** CAUTION"
7300 PRINT "  SMALL DISPLACEMENT =", TA
7310 PRINT "   ACCURACY CHECK RECOMMENDED *********"
7320 PRINT
7999 RETURN
8000 REM ELEMENT FORCES
8010 PRINT
8020 PRINT "ELEMENT SUMMARY"
8030 PRINT "ELE N1  N2     A          E        L      IL       F
STR"
8040 PRINT
8050 IL = 4
8060 FOR I = 1 TO N3
8070 REM ELEMENT STIFFNESS
8080 GOSUB 3228
8090 IN = E1(I) * 2
8100 ITN = E2(I) * 2
8110 I1 = E1(I)
8120 I2 = E2(I)
8130 II(1) = IN - 1
8140 II(2) = IN
8150 II(3) = ITN - 1
8160 II(4) = ITN
8170 FOR L = 1 TO 4
8180 SUM = 0
8190 FOR K = 1 TO 4
8200 JJ = II(K)
8210 SUM = SUM + EE(L, K) * Q(JJ)
8220 NEXT K
8230 SE(L) = SUM
8240 NEXT L
8250 EF(1) = COS(TH) * SE(1) + SE(2) * SIN(TH)
8260 EF(2) = -SIN(TH) * SE(1) + COS(TH) * SE(2)
8270 EF(3) = COS(TH) * SE(3) + SIN(TH) * SE(4)
8280 EF(4) = -SIN(TH) * SE(3) + COS(TH) * SE(4)
8290 IM = M(I)
8300 AR = A(IM)
8310 EF3! = EF(3)
8320 ST! = EF(3) / AR
8330 UL = XL / LF
8340 F$ = "## ## ## ###.#    ########.#    ####.#    ####.#
######.# ######.#"
8350 PRINT USING F$; I; I1; I2; AR; YM; XL; UL; EF3!; ST!
8360 IL = IL + 1
8370 IF IL < N6 GOTO 8450
8380 INPUT ; "TOUCH ANY KEY TO CONTINUE", Z$
8390 PRINT
8400 PRINT T$
8410 PRINT
8420 PRINT "ELE N1  N2     A         E      L       IL      F      STR"
8430 PRINT
8440 IL = 4
8450 NEXT I
8499 RETURN
9000 REM
9010 REM USER DEFINED SUBROUTINES
9020 REM
20000 REM
20000 REM
20010 REM TITLE FOR PROBLEM
20020 DATA "WARREN TRUSS EXAMPLE"
20030 REM NUMBER OF NODES, LENGTH FACTOR
20040 DATA 5,12
20050 REM X,Y COORDINATES (UNITS = FEET)
20060 DATA 0,0
20070 DATA 5,10
20080 DATA 10,0
20090 DATA 15,10
20100 DATA 20,0
20110 REM MATERIALS- NUMBER
20120 DATA 1
20130 REM MATERIALS VALUES: AREA,MODULUS
(UNITS=INCHES)
20140 DATA 10,36000000
20150 REM ELEMENTS- NUMBER
20160 DATA 7
20170 REM NODE1, NODE2, MATERIAL TYPE  OF ELEMENT
20180 DATA 1,2,1
20190 DATA 1,3,1
20200 DATA 2,3,1
20210 DATA 2,4,1
20220 DATA 3,4,1
20230 DATA 3,5,1
20240 DATA 4,5,1
20250 REM BOUNDARY CONDITIONS - NUMBER
20260 DATA 3
20270 REM BC'S NODE NUMBER, DIRECTION (X=1, 2=Y)
20280 DATA 1,1
20290 DATA 1,2
20300 DATA 5,2
20310 REM     LOADS - NUMBER
20320 DATA 2
20330 REM NODE, DIRECTION,     LOAD VALUE
20340 DATA 2,2,-100000
20350 DATA 4,2,-50000
```

Programmer's Guide to TF2DFE

This program is written in what is traditionally considered "10 statement programming." In the context of computers today that implies that it is written in the simplest matter possible to make the code brief and transportable to various software and hardware platforms. The code was produced using the Microsoft (a registered trademark) company BASIC software standard in the Windows 95(a registered trademark) installation disk. (Note that the mention of trade names is not an endorsement by the publisher or the author.) This section details the implementation of the code in this software to assist in projects which might be undertaken to extend the capabilities of the code.

The statements of BASIC used in the code are as follows:

REM - a note or comment, produces no action.

READ - a statement to read data from a data statement using variables in the arguments (items which follow the read). Reading is sequential in the data statements.

INPUT - a statement to input data from the keyboard. Used in this program to create a pause by reading information from the keyboard.

DATA - statements which contain data for the READ statement.

DIM - storage locations for data which can be grouped as to use, for example X coordinates for all nodes.

PRINT - statement to print variables to the screen.

PRINT USING - same as print but uses a defined format for the variables.

FOR — NEXT - a pair of statements used to execute a series of program lines repeatively and to increment the index of the FOR statement with each subsequent pass through the code.

IF arg GOTO program number - a condition that if the argument is true the code branches to the specified program statement number, otherwise proceeds with the next statement.

Example: IF A(I,J)=0 GOTO 1320.

IF THEN used equivalently

GOSUB - branch to code which is known as a subroutine.

RETURN - return to next statement after GOSUB when subroutine is complete.

END - end program execution

STOP - used equivalently

Functions called from the program

ABS - return absolute (positive) value of argument.

ATN - returns the tangent in radians for the argument.

COS - returns the cosine in radians of the argument.

DEFDBL A-Z - uses double precision for variables to assure accurate calculations as numbers get small

SIN - returns the sine of the argument.

SQR - returns the square root of the argument.

Dictionary of Major Variables for the Program

A( ) - Area by material number
AC - Accuracy check = 0.00001, used to check if small numbers may influence accuracy of results, rule of thumb for use with single precision calculations
AR - Area of an element

B( ) - External load equation number
B - Input value for boundary condition node number, converted to B( )
BC( ) - Equation number of fixed degrees of freedom
BV - a specified displacement boundary value, coded as zero in this code

C - Cosine of an element, measured from node 1 to node 2
CC - Cosine squared
CS - Cosine times sine

D - Input value for direction
DX - Delta x of an element (x vector from node 1 to node 2)
DY - Delta y of an element (y vector of an element)

EE(,) - Element stiffness matrix
EF( ) - Element forces in element coordinates
E1( ) - First node of an element
E2( ) - Second node of an element

F$ - Format to print variables.

G(,) - Global Stiffness Matrix

IC - Counter of number of lines printed
II( ) - Element to Global Coordinates
IG - A position in the global stiffness matrix
IM - Material number
IN - Element to Global Stiffness equation number
ITN - Element to Global Stiffness equation number for node 2
I1 - First node of an element
I2 - Second node of an element

JG - A position in the global stiffness matrix

KP - A position in the global stiffness matrix

LF - Length factor to convert length to consistent units, usually 12 to convert coordinates from feet to inches

LP - A position in the global stiffness matrix

M( ) - Material type for an element
MM - Bandwidth of global stiffness matrix

NN - Number of degrees of freedom
N1 - Number of nodes in problem
N2 - Number of materials in problem
N3 - Number of elements in problem
N4 - Number of boundary conditions in problem
N5 - Number of loads in problem
N6 - Number of lines on user's computer screen
N10 - Maximum number of nodes (=25)
N20 - Maximum number of materials (=10)
N30 - Maximum number of elements (=100)
N40 - Maximum number of boundary conditions (=20)
N50 - Maximum number of global degrees of freedom (=80)
N59 - Maximum number of loads (=100)

P( ) - Value of an input load

Q( ) - Load vector of assembled (global) loads

S - Sine of element, measured from node 1 to node 2
SE( ) - Element forces in global coordinates
SS - Sine squared
ST - Stress

T$ - User's input title for problem
TA - Test accuracy computed by program
TH - Theta value of an element to the global axes, in radians
THC - Value to calculate tangent (DY/DX)

X( ) - X coordinate of node
XL - Length of an element

Y( ) - Y coordinate of node
YM - Young's Modulus

Z$ - a temporary variable to read value from keyboard, used as a pause

The Fun Projects in the appendix suggest some easy and some more sophisticated additions.

Chapter 7 - Strength of Materials

Part of engineering involves knowing how strong a material can be. This is called the material strength. Material strength is measured by the uniform load a material will hold when it fails.

The objective of Strength of Materials is to understand the development of relationships between loads applied to a nonrigid body and the internal forces induced in the body. For the beginning engineering student, four Strength of Materials principles provide a first lesson:

1. The effects of load on a material, stress

2. Deformation of a material, strain

3. The relationship between stress and strain in a material, Hooke's Law

4. The relation of load to bending

Materials science is related to the fundamental properties of materials when subjected to loads. That is, what happens to all materials when subjected to forces. Material properties which are common to all materials are:

1. Linear
2. Elastic
3. Proportional Limit
4. Yielding
5. Plasticity
6. Ductile
7. Brittle

In the following sections each of these principles is explained.

7.1 Properties of Linear and Elastic

We began a spring experiment in Chapter 6 to demonstrate the Spring Law. The experiment placed a load on a spring and notes that a displacement resulted. If we continue with our example by applying a second load, say of equal value, will we get a proportional value of deflection (twice the load, twice the displacement)? For small loads and small displacements it is assumed that load and displacement will be proportional and can be related by a constant. The material property, linear, is defined by this proportional relationship. A spring could be stated as being linear since a proportional value of load produces a proportional value of displacement, e.g. twice the load produces twice the displacement, half the load produces half the displacement, etc. Note that linear data will plot as a straight line when data is presented graphically.

Linear: Proportional

Continuing with our experiment, one load is removed. What do we expect to happen? With a spring, it is reasonable to expect that the spring will unload to the same displacement as when the first load was applied. When material displacement is recoverable (load and unload on the same path) it is considered to be elastic.

Elastic: Load and Unload on the same path = Recoverable

7.2 Principles of stress and strain

Normal strain is used as an expression of structural deformation which eliminates the dependence on geometric shape. This is accomplished by dividing the deformation by the length. In Figure 7.1 a round bar is shown subjected to a force, P. With P applied a deformation may be measured over a selected portion of the bar, y.

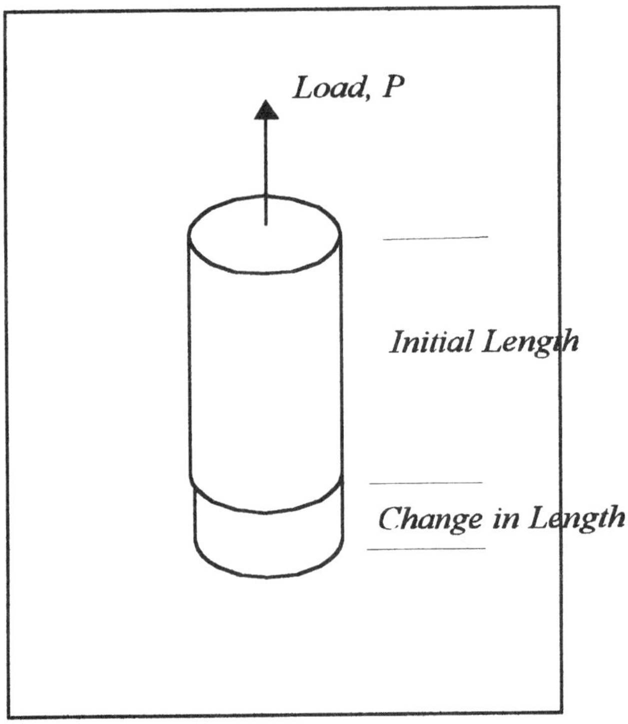

Figure 7.1 Bar under Tensile Load

The deformation may be divided by the length of the bar over which the measurement was taken, L, to find the value of strain, or:

Normal Strain:

$$\varepsilon = y/L$$

Where:

ε is defined as normal strain
y is deformation
L is the length over which the deformation is measured.

Strain, like force, is commonly referred to as tension or compression depending on the direction of force. Tensile strain is associated with an elongation of the material, whereas compressive strain is associated with a shortening of the material.

Normal stress may be introduced as a variable which provides a property which is based on a unit area of a structure, it is found by normalizing the load, P, with the cross sectional area, or:

Normal Stress:
$$\sigma = P/A$$

Where:

σ is defined as normal stress
P = load applied to the member
A = the cross sectional area of the member

Like strain, normal stress is referred to as tension or compression.

7.3 Stress- strain diagrams and Hooke's Law

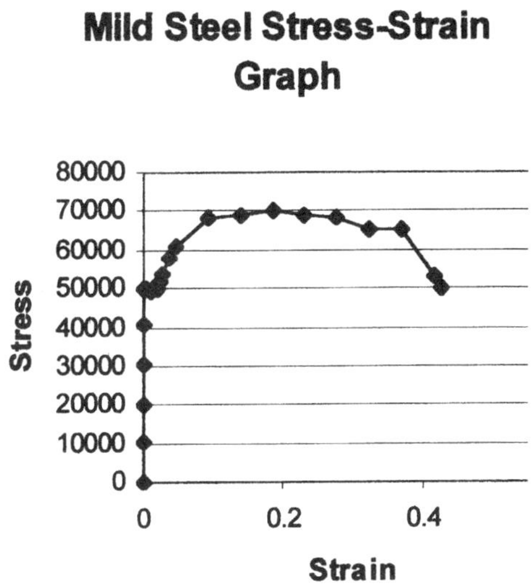

Fig. 7.2a - Typical stress- strain diagram for Mild Steel

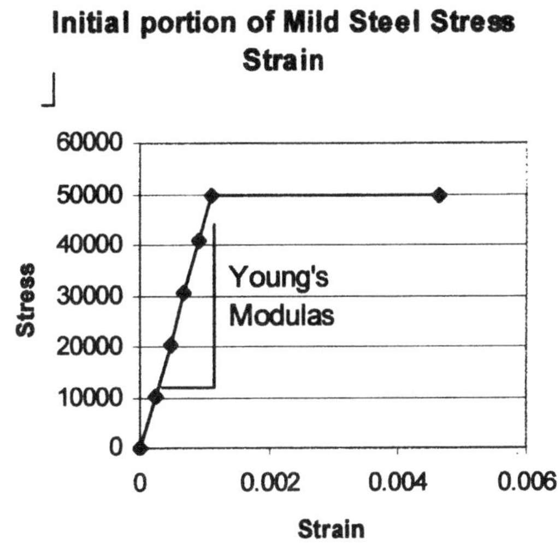

Figure 7.2b Expanded scale stress-strain graph

With the fundamental quantities of stress and strain defined, a graphical visualization of the stress-strain relationship, which can be obtained in the laboratory, can be drawn. Imagine a simple round bar subjected to a tensile load which increases gradually. If simultaneous values of load, P, and elongation, are obtained a stress- strain diagram can be constructed. As noted in the previous definitions, stress would be obtained from the loads by dividing by the area. A typical test specimen of a ½" bar would have an area of 0.196 square inches. A typical load in the test would be 10000 pounds, for a stress of 10000/0.196 or 51,000 psi (pounds per square inch). A typical diagram for low carbon steel is shown in Figures 7.2a and 7.2b.

In the expanded scale plot, Figure 7.2b, it can be seen that the first part of the data is generally linear. The material property of linear generally means that stress is proportional to strain. This relationship can be expressed as a straight line plotted on the stress-strain graph, or:

$$\sigma = E\,\varepsilon$$

This relationship is known as Hooke's Law. The constant E, is the initial slope of the stress- strain graph, and is known as the Young's Modulus.

It is from Hooke's Law that a stiffness constant can readily be found for trusses. We know that trusses carry only normal load. A normal stress is P/A. For normal strain, the deflection is also referred to as y, and is divided by the original length, or y/L. Rearranging these relationships yields:

$$\sigma = E\,\varepsilon$$
$$P/A = E * y/L$$
$$AE/L * y = P,$$

This is in the form of the Spring Law, K y = P, so

$$K = AE/L$$

7.4 More Material Characteristics

Certain characteristics of the stress- strain diagram have been defined and named. These definitions will be important in making comparisons between properties of materials. Observe that the initial portion of the diagram is straight (linear - stress is proportional to strain) followed by portions which are curved (non-linear). The maximum stress for which the plotted data remains straight is defined as the Proportional Limit. More formally, the Proportional Limit is the maximum stress for which Hooke's Law applies.

The Yield Point Stress is the first stress for which there is a marked increase in strain with little or no increase in stress.

Yielding of material is the flow of the material without failure. This flow of materials is known as plasticity. Large strains can be associated with this flow. Strains encountered while loading along this upper portion of the stress- strain curve are plastic strains.

Materials which can sustain large strains before failure are said to be ductile. Materials which fail after small strains are said to be brittle.

The way that materials fail and their stress-strain characteristics can be used in scale models to simulate failures. This type of modeling will be covered in Chapter 9.

7.5 Stresses in bending

Loads in the span of a structure (i.e between supports) cause bending. Bending is the deflection away from the original shape of the structure. Two important concepts of truss design used throughout this book are: (1) loads are only applied at nodes, (2) loads are only created along the elements of the truss. With loads in the middle of a truss, (not at a node), some new concepts arise.

Moments can exist inside structures under loads which fall in the midspan of a member. It will be easiest to understand this with an example. The simplest example is a beam supported at two points, Figure 7.3a.

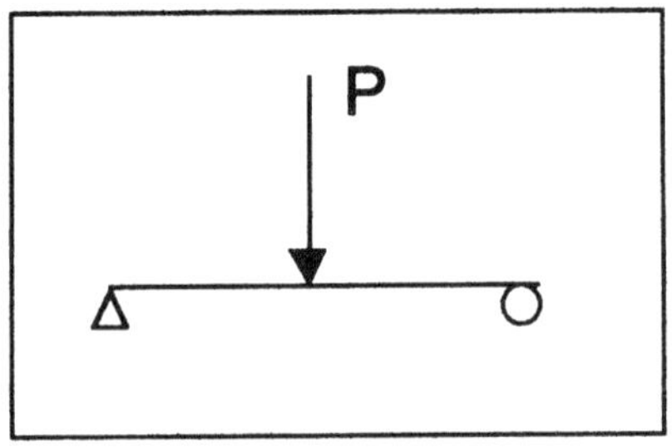

Figure 7.3a - Simple beam

First imagine a single force in the middle of the span, and its free body:

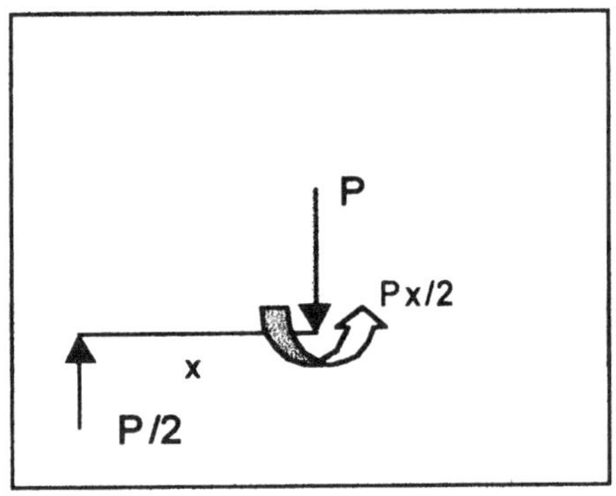

Figure 7.3b Simple beam and free body

The free body contains forces which balance. By using the sum of forces in the y direction with the entire structure, the reaction at the support is easily found as P/2. On the freebody, the sum of moments about the point used to take the freebody, (the right end) must equal zero, or P/2 * x (where x is the distance from the support to the point of the freebody) must be the moment on the section. This moment would be denoted as the value of M to be used in the calculation of stresses due to bending with a load at the midpoint of a beam.

We can introduce the equation for stress in bending as:

$$\sigma = Mc/I$$

Where:

σ = stress caused by bending
M = the moment on the section
c = the distance from the centroid to the outside fiber
I = the moment of inertia of the cross section.

The equation is derived from classic equations which are beyond the scope of this book.

A harder problem would be a distributed uniform load, w, on a simply supported beam:

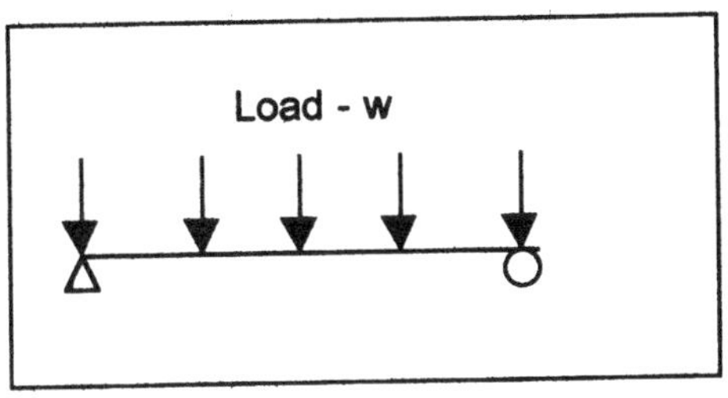

Figure 7.4 Simple Beam with a Distributed Load

The reaction can be found fairly easily as wl/2. Note that the moment caused by the distributed force will be the force on the freebody, wx, times the moment arm of the force, in this case the distance from the center of the distributed load, x/2. Distributed loads may always be replaced by an equivalent load which is the value of the load acting through the geometric centroid of the load. The free body now contains both the reaction and the distributed load in the free body. Thus, forces must be adjusted to account for both. The full moment on the freebody section is found using the sum of forces as before and is given by $M= -wlx/2+wx^2/2$.

With multiple loads, the effects can be added and subtracted (using the correct sign)to find a total load. This has a somewhat fancy name: Superposition.

More information can be found in textbooks used to study engineering. The terms and ideas of bending are introduced here as an introduction to understand various loading conditions.

Table 8.1 - Selected Truss Failures

Date	Bridge	Location	Comment
May 11, 1858	44 foot wooden	Sauquoit Creek	Collapse from two trains
1873	Wood River Junction	Rhode Island	Upstream dam failed from heavy rains and washed out bridge. Train going to fast to stop.
1876	Iron Howe Truss	Ashtabula	Failure of bridge
Dec. 28, 1879	Tay Bridge	Scotland	Bridge fails under train
1881		Sunbury Ohio	Bridge fails under train
June 12, 1882	Camp Creek		Rain leads to collapse
Oct. 17, 1884	Little Miami River	Batavia Ohio	Bridge fails under train
Spring 1887	Busey Bridge (Iron bridge)	Forest Hills station near Boston	Loose bolts and nuts discovered below bridge before failure 1887 accounts for 21 bridge failures
1893	Indianapolis and St. Louis Road	Hillsboro, Ohio	Bridge Washout
Aug. 13, 1900	Carrolton Viaduct	Gwynn's Falls, MD	Side collision of bridge
1907	Boston and Albany Bridge	Chester, MA	
Aug. 8, 1907	Bridge over Loire River	Angers, France	See text
September 27, 1923	Cole Creek	Lockett, WY	Flash Flood weakens supports
July 18, 1929	Rock Island Railroad bridge	Stratton, CO	Flash flood sweeps away bridge
Aug. 31, 1933	Anacostia River Bridge	4 miles north of Washington DC	Settlement of pier causes derailment
Aug 29, 1933	Southern Pacific Line	Tucumcari, NM	Settlement of pier and derailment due to rains
April 6, 1950		Rio de Janeiro, Brazil	Train wreck by bridge collapse
Dec. 24, 1953		Waiouri, New Zealand	Train plunges through bridge
Sept. 2, 1956		Mahbubnagar, India	Bridge collapses beneath train

Chapter 8 - Structural Stability

A structure will fail if too much load is applied. The first possible way to fail (the least load) is the way the structure will fail. This basic idea is known as the Least Energy Theorem.

In Chapter 7, material strengths and material failures were discussed in detail. In this chapter, other structural failure possibilities are discussed. These possibilities relate to the structure, or structural stability.

8.1 Conditional Stability

Structures may be stable for some loading conditions, but unstable for others. For example, if a single truss element hangs vertically from a pin, the truss would be stable for a vertical load, but would not be stable for a load perpendicular to the truss, as there is no stiffness in that direction. This is known as conditional stability.

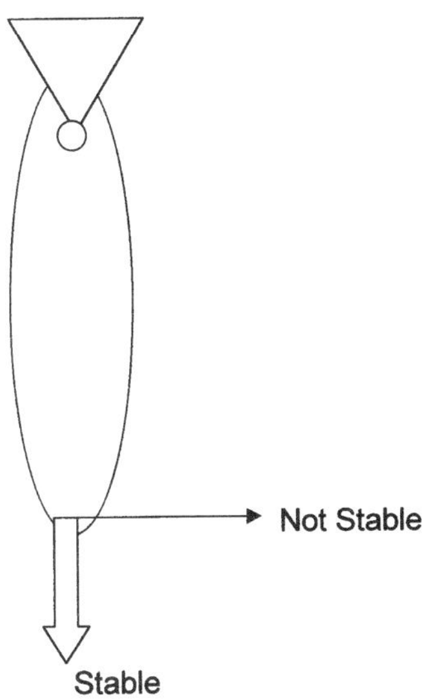

Figure 8.1 Conditional Stability

Trusses made up of triangles are generally stable, provided the supports, or boundary conditions provide stability. If r = the number of reactions, (or boundary conditions), and j = the number of joints in a truss, then a truss will definitely be unstable if the number of members is less than 2j-r. Unfortunately, a definite test to assure stability is not as straight forward.

8.2 Failure of joints and connections

Structures can fail because a connection or joint is not strong enough to bear the loads. In essence this is a material failure of the joint materials. What may not be obvious in the construction of joints, such as in models, is that materials such as glue are not strong in tension. Joint design in real structures is an important aspect of the design.

There is another type of force known as shear force which requires a material to slide across itself (like sand paper). A material property, known as shear strength, would be used to compare loading conditions to failure conditions. If the shear loading in a structure exceeded the shear strength of the material, the structure would fail.

Materials are not as strong in shear as in compression. Say for example, concrete, a material which is strong in compression, is only about 1/3 as strong in shear, and about 1/10 as strong in tension. Thus, a loading condition that causes shear can make a difference in the ability of joints to withstand loads.

Simple truss analysis generally assumes connections may be treated as pins for analysis purposes relative to the truss. The connections need to be carefully checked to assure that moments and shears are not present which may exceed the strength of the joint.

The design of structures and models requires the consideration of many different conditions. Some additional conditions follow in the subsequent sections.

8.3 Buckling

Buckling is the creation of large displacements lateral to a member subjected to compression loads.

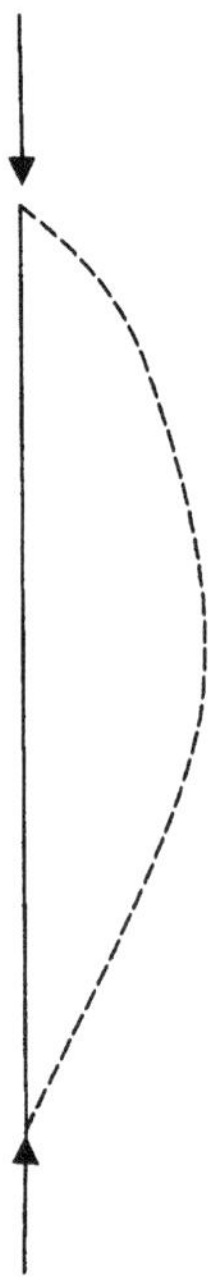

Figure 8.2 - Displacement from buckling

Once the displacement reaches the critical point, the buckling load, the results are a sudden and spectacular collapse. This can be demonstrated at home. Take a soda straw and press on it from the top. A little bending will occur at first but suddenly the straw will buckle and collapse under almost no load.

Buckling was initially studied by Leonard Euler, a Swiss mathematician, who lived from 1707 to 1783. He used the principles of bending to arrive at the critical load necessary to buckle a pinned column (like a truss member). His relationship is well known as the Euler Equation:

$$P_{critical} = \frac{\pi^2 EI}{L^2}$$

Where: P = critical load
E = Young's Modulas
I = Moment of Inertia
L = Length of column

Note that this condition may be more critical than material strengths when there are long slender members in a truss.

8.4 Foundations, Floods, Landslides, and Earthquakes

There are many components to a structure, all of which must be stable for the structure to be stable. The foundation of the structure, where it sits on or into the earth must be stable. If the foundation yields, or gives way, the structure may also fail.

Floods have been the cause of bridge failures in some cases. The higher waters can cause unexpected loads on the bridge supports, or can erode away the foundation of the bridge. Chapter 4 showed a design of the bridge foundations in river beds to insure that water can not wash away the foundations.

Hillsides can slide and fail a structure, or hit a structure and create loads much larger than expected in the design.

Large loads caused by an earthquake can fail a structure. In designs of today, wind, and other natural loading conditions are suggested in standards to assure safe structures. These loads have been developed from historical cases and research and were not used, or well known in early designs.

8.5 Impacts from Trains

A member of a bridge could fail from the load or impact of a train. Derailments may cause a train to strike a bridge member and fail it and thus cause the entire bridge to fail. Historically there have been bridge failures for a number of reasons.

8.6 Selected Historical cases

In the United States, as well as in other parts of the world, the development of railroads led to improvements in design. Some of these design improvements came about as the result of an accident. As the number of miles of track increased, the amount of data increased from more usage and more total miles of record.

Some of the early trestles were too lightly constructed for the heavier and faster stock. In 1887 alone there were 21 bridge collapses. Some of these known failures are shown in Table 8.1.

In 1829-1853 trains ran slowly on the rails, usually 10-18 miles per hour. There were not many trains in operation and there was virtually no night travel. In 1835 approximately 1100 miles of track existed. By 1850 the miles of track had increased to almost 10,000 miles. Following 1850, the total miles of track would increase rapidly throughout the nation.

As track increased, so did railroad bridges. In 1840-1870 the standard bridge was the Howe truss. This truss, a timber truss primarily, was vulnerable to fire, decay, and floods. Derailments occurred which in many cases damaged the bridge.

In 1870, with the widespread use of iron for building equipment, the rolling stock available to railroads would rapidly increase. Trains would become larger and heavier, with more frequent trips.

Historical accounts of failures provide insight into the developments of modern designs and design standards. Two such accounts of the actual failure of truss bridges are presented here for the potential interest of the reader.

The Tay Bridge Disaster

Perhaps one of the most famous railroad bridge failures is the Tay Bridge Disaster at the Firth of Tay near Dundee Scotland. This bridge, built between 1871-1877, was the longest crossing over water of its time at over 2 miles. The site chosen for the crossing was not the shortest possible distance but appeared to have the most favorable foundation conditions.

The chief designer was Thomas Bouch of the Edinburgh & Northern Railway. The initial design called for girders and deck trusses from each shore. The main shipping channel crossing was designed as 13 lattice through trusses, each 200 feet in length, with a clearance height of 88 feet over the shipping channel. These trusses were referred to as the high girders.

The foundations for the bridge needed to be dug by hand inside large caissons (caissons are large cylinders extending below water level to the rivers bottom for access by the workers). The work was grueling and difficult. Once the caisson had reached rock it was lined with brick and filled with concrete. There were problems reaching rock in the foundations. The original soundings had located a gravel layer which laid above the rock level. Thus, depths of the foundation were increased during construction. In addition, the installation of the caissons had problems. With the initial design one side of the caisson would hit rock first and subsequent digging would result in tipping of the caisson. The design was shifted to a double barrel caisson, which was more stable. As construction progressed and costs mounted, the foundations were switched to a 31 foot diameter caisson which could be built on shore, floated into location, sunk to the floor and filled with concrete. The number of supporting columns for the high trusses, each 86 feet in height, was changed from 8 to 6 to meet the reduced foundation dimensions. The span of the trusses was changed to 245 feet to eliminate one pier. In addition, a more easily assembled header was designed, not the original one piece header. All of these design changes were an effort to stop escalating costs of the construction.

The high girders, the lattice trusses, were fabricated on shore. The 245 foot long trusses were 18 feet across and 36 feet high. Each truss required 4 weeks of labor, 18,000 rivets and weighed 190 tons. All of this at a time when the development of iron materials was not in pace with the needs of a rapidly growing bridge industry. The trusses were lifted into place using hydraulic lifts. The last rail of the bridge was finally placed on September 27, 1877. Costs for the bridge totaled 320,000 pounds Sterling, or an overrun of 48% over the initial budget.

The bridge was inspected by government inspectors using increasing numbers of locomotives and cars. After extensive testing the bridge was approved for traffic on March 31, 1878 with a future inspection planned. The speed limit was set at 25 miles per hour.

A crew of seven men was assigned as a maintenance crew to check bolts on the bridge. In September of 1878, the wind ties, braces for the high girders, were found to be loose on the piers. Bolts reportedly fell from the bridge as trains passed. By this time trips over the bridge were becoming routine and speeds of 40 miles per hour were common. Passengers reported a rattling sound and a prancing motion as the train crossed through the main trusses.

December 28, 1879 saw a sudden storm developing at around 5:00 pm with fierce winds of approximately 75 miles per hour. Two eyewitnesses reported cyclone like columns. The 7:00 pm train arrived to cross the bridge with one locomotive, a tender, a third class car, a first class car, 2 more third class cars, a second class car and a van. This train of 225 feet weighted 114 tons. As the train crossed through the ship channel trusses (the high girders) all 13 trusses and the entire train crashed into the water below.

Following the disaster, a board of inquiry was convened. Testimony from various witnesses revealed problems in various areas. It was found that there were difficulties in forming the cast iron members. Soft spots discovered in the casting had been filled and painted over. Inspectors had located cracks in the columns after assembly. Workers on the columns produced oversized and unround bolt holes of uneven diameters on the columns. Bolts had been reported to fall off the bridge. The design of the bridge was done by a designer unfamiliar with the use of iron; wind loadings were not considered in the design; and major changes were made during the construction to save on expenses. Trains routinely traveled at speeds exceeding the inspectors recommended maximum speed.

Better standards and higher factors of safety in design resulted from suggested inquiry recommendations. (The factor of safety is the ratio of the design load to the failure load. A factor of safety of 2 would mean that a structure should not fail until the load is two times the load used in design. The ratio improves as it increases over 1.0). The Tay Bridge was rebuilt and today is listed as the second longest steel truss bridge in the world.

Old Railway Bridge in France

On August 1, 1908 a passenger train crossing a Lattice Truss bridge over the Loire River near Angers, France broke through the floor of the bridge. Part of the train fell into the stream and 27 persons were killed. The bridge was built in 1876 and consisted of spans of approximately 130 feet. The floor beams of the bridge consisted of plate girders. Train weights had increased considerably since the design and construction of the bridge.

A train consisting of one locomotive, a tender, a baggage car, four third class coaches, one first[t] class coach, two second class coaches, and a caboose had just entered the first pier of the main channel truss when the failure occurred. The floor tore loose dropping the locomotive, tender, baggage car, and first coach. The locomotive in this case was actually light, weighing only 45 tons. The failure was believed to have been caused by the derailment of the train before entering the truss. The train impacted the bridge which subsequently failed. In this bridge no rail expansion devices were provided between the bridge and the approach rails. The bridge had no guard rails in spite of an earlier accident in 1903, when a train had derailed in the same span, tore through part of the lattice of the truss, and did considerable damage to the floor of the bridge.

Chapter 9 - Modeling Failures

Chapter 5 discussed making scaled geometric models of full size structures. Models are usually made from materials different from the full sized structures they represent. For example, structures may be made of steel, whereas models may be made from plastic shapes available from supply houses.

If model structures are built with the intention of testing to failure, either for fun or to simulate a historic case, some additional scaling relationships need to be considered. These additional relationships are known as the Laws of Similitude.

9.1 Laws of Similitude

The laws of similitude make small objects act like larger, prototype (full sized) objects. It is obvious that objects which are at a smaller geometric scale will weigh less, since they have less material. So loads will be adjusted to the scale of the model if the same materials are used, or materials of the same density.

The key to understanding failures at model scale is that the strain at failure must be the same in both model scale and full scale to properly measure the failure. This is true since strain is dimensionless, totally independent of size and shape. Recall that strain is the deflection divided by the initial length. Length is a geometric parameter and was scaled to make the model. The deflection needs to be scaled so that the strain will remain on a unit basis. Figure 9.1 shows the desired failure condition for both the model and the prototype (full sized) structure. The strain at failure is the same for both. The loads and thus the stress in the model will be much lower.

Imagine a shrinking machine. We turn on the shrinking ray and type in a scale of say 1/10. There is a lot of noise, and then a little window pops up which says, enter parameters for the scale, for example, scale by length or area? In the following sections, the different scaling relationships are discussed. Text, or units, can be used in the strain equation to demonstrate this. When parameters are used at scale the word scale is substituted for the parameter and the equation solved:

Strain = "Defl"/"Length"
Strain = Scale/Scale =1
Strain same in model and prototype

Several cases need to be considered: Case 1. Structures which are made up of continuous materials like soil foundations, and concrete in foundations and bridge abutments (end supports), and 2. Structures, like trusses which are made of individual elements which can be scaled.

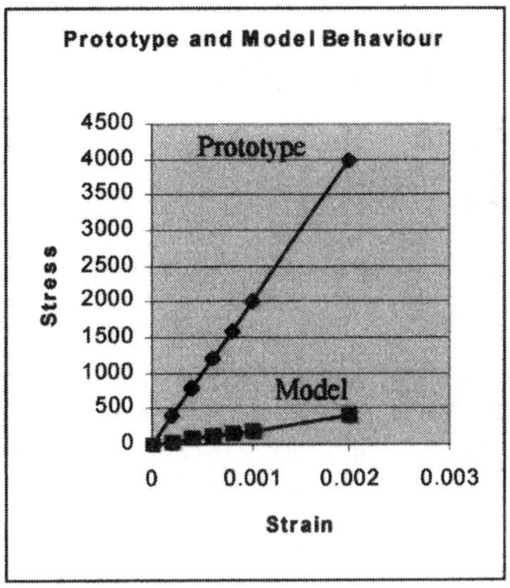

Figure 9.1 General relation of modeling materials to failure.

9.1.1 Concrete and Soils

The first case considered , Case I, relates concrete and soils which act on a unit area basis when stress is applied. Unlike stresses in members like in trusses, we have to base our models on what happens on a square inch basis. Length is scaled by the size change. If we use materials in the model that are a similar density to the field case, the loads on a square inch basis are scaled in the structure by the change in the size of the structure. Thus, stresses are represented at scale.

The other parameter to be considered is the stiffness constant, E. If E were not scaled, the deflection would be very small, due to the division by a large number. The deflection, and thus the strain, would not be scaled. Therefore, the modulus must also be scaled.

Changing the strength and the stiffness to the same ratio may be difficult. In addition consider that some historical cases of failure may be the result of foundation failure in full scale, so it is the foundation material, and not the truss, which must be modeled accurately using similitude.

For convenience, all similitude cases are summarized in Table 9.1. This provides a ready reference for modelers and for comparison of the various approaches which follow.

Table 9.1 Summary of Similitude Modeling Cases

Parameter	Case I - Continuous	Case II - Trusses - Length and area scaled	Case III - Trusses - Length and dimensions scaled	Case IV - Trusses - Length, area and Materials scaled	Case V - Trusses - Length and area scaled, Material - Fractional Similitude
Area, A	Unit = 1	Scale	Scale*Scale	Scale*Scale	Scale*Scale
Length, L	Scale	Scale	Scale	Scale	Scale
Modulas, E	Scale	Model=Field	Model=Field	Scale	E-Ratio
Deflection, y	Scale	Scale	Scale	Scale	Scale
Load, P	Scaled with density of model = density of field	Scale	Scale*Scale	Scale*Scale*Sca le	Scale*Scale*E-ratio
Stress	Scale	Model=Field	Model=Field	Scale	E-Ratio
Failure Stress	Scale	Model=Field	Model=Field	Scale	E-Ratio

Notes:

1. Case II-
   Controlling Equation: y=P — Best Similitude relation - difficult to size members
2. Case III-
   Controlling Equation: Scale*y=P — Field materials not available in model size
3. Case IV-
   Controlling Equation: Scale*Scale*y=P — Difficult to scale material properties
4. Case V-
   Controlling Equation: Scale*E-ratio*y=P — Best approximation for model materials

9.1.2 Trusses

When models are made from real structures, the scale may be based on length - all dimensions. Changing all dimensions will make the structure look in scale. The scale could also be based on a scale change to the area of the members.

The area is also a geometric parameter, but note that for rectangular elements, both dimensions are scaled. For example, suppose a 10" x 10" member is to be sized to a 1/10 scale. The initial area is 100 square inches so the desired area is 10 square inches. If all dimensions are scaled the cross section becomes 1 inch by 1 inch or 1 square inch, a scale of 1/100. Trusses respond to stress and strain based on the area. So for geometric models we would choose to scale by length, whereas for failure models we would choose to scale based on area. This is going to require some decisions when making a final model.

Case II Trusses - Scale by length and area

Once a scale is established for the length, the area of truss members could be reduced by the same scale. That is if the scale was 1/87, the lengths would be reduced by this scale, the area would be reduced by the same scale. New dimensions would be established from the area. Materials would be identical (steel to steel) or similar in the model and the field (the stiffness E would be the same). In this approach the following would then apply.

First, recall the Spring Law,

$$\frac{AE}{L} y = P$$

Substitutions will be made for items which will be used at scale to resolve the scale relationships.

$$\frac{\text{"A" "E"}}{\text{"L"}} \text{"y"} = \text{"P"}$$

$$\frac{\text{Scale}(1)}{\text{Scale}} \text{"y"} = \text{"P"}$$

$$\text{"y"} = \text{"P"}$$

and if displacements, y, are in scale, the loads, P, will be in scale, or conversely if P is to scale then displacements are to scale. The loads would be held to scale by using materials of the same density and construction. This approach certainly has appeal, but note that models, such as train models are not likely to be manufactured at scale with regard to their weight. Models will generally be taken to failure with weights added to the structure.

Stresses calculated at model scale would be

$$\text{Stress} = \text{"P"}/\text{"A"}$$

$$\text{Stress} = \text{Scale}/\text{Scale} = 1$$

So stresses in the model would be equal to stresses in the field and the failure strength of the model material could be the same as in the field. This creates a fairly easy modeling criteria but the visual appearance will be distorted and nearly all members will need to be custom made as the dimensions will be odd values.

Case III Scale by length and dimensions

In this modeling approach the length of the model and the member dimensions are reduced by the scale. As rectangular dimensions are multiplied twice (area = base * height) and diameters are squared (area = 3.14*diameter*diameter/4) multiplying the cross section dimensions each by the scale, or the diameter by the scale will have the same effect. That is, area is factored by scale*scale. The stiffness is the same in the model as in the field.

The deflection is assumed to be to scale in order to find a relationship for conversion of loads. Note the assumption that the deflection is to scale is an approximation.

Substituting into the Spring Law,

$$\frac{(\text{Scale*Scale})(1)}{\text{Scale}} \text{Scale} = \text{"P"}$$

"P" = Scale*Scale

That is, loads used in the model would be converted to field loads using the product of Scale*Scale (e.g. if a model scale is 1/48, model loads would be multiplied by 2204 = 48*48 to convert to field loads).

Stresses would be found as:

Scale*Scale/(Scale*Scale) = 1

So, as before, stresses in the model and field would be identical. Failure strengths would also be identical.

Case IV Length, area, and material scaled

With this approach, the material stiffness, E, is also attempted as a scale. The advantage of this approach is that the model can use materials that have a lower E, which will produce larger displacements, which are easier to measure and fun to see.

Proceeding as before,

$$\frac{\text{"A" "E"}}{\text{"L"}} \text{"y"} = \text{"P"}$$

$$\frac{\text{Scale*Scale*Scale}}{\text{Scale}} \text{Scale} = \text{"P"}$$

"P" = "Scale*Scale*Scale

Stresses are readily found by substitution

Stress = Scale*Scale*Scale/Scale*Scale

Stress = Scale

So the stresses in the model are in scale relative to the field case. For failure to occur at the proper stress condition, the failure strength must also be scaled.

Case V - Scale length, area, and use modeling materials as available

More typically in modeling, the length scale will be determined as a geometric model, dimensions will be scaled, different construction techniques from the full scale structure will be used and materials readily available in hobby stores will be used in the model. This requires that the final failure scale be estimated.

The same approach as before is used. In this case the material property, E, is used at some ratio of model to field, not necessarily at the scale of the model. This term is defined as E-Ratio.

$$\frac{\text{"A" "E"}}{\text{"L"}} \text{"y"} = \text{"P"}$$

$$\frac{\text{Scale*Scale*E-Ratio}}{\text{Scale}} = \text{"Scale"} = \text{"P"}$$

Scale*Scale*E-Ratio = "P"

Stresses are found by substitution as

Stress = E-Ratio

So as in Case IV, a conversion needs to be made between the model and the field and the failure strength needs to be adjusted by E-Ratio.

Due to the nature of multiplying fractions to estimate the final scale relationship, this approach has been designated as Fractional Similitude. It is suggested in this text for modeler's as a good estimate for models which are practical to construct.

This approximate approach is used as an example which follows in the student workbook. The Fractional Similitude relations for this case are included as an example. As a project, a 202 foot Howe truss was modeled and tested to failure. The original drawing for the truss was drawn as 1/4" equals a foot. Wood members in the truss are generally 10" x 10". The final model scale was decided to be 1/48 so the original drawing would represent the final size of the model and the drawing was used as a cutting guide. Thus length was scaled by 48 and the final bridge was 202/48= 4.2 feet long.

Members were scaled based on length so that the visual appearance of the final bridge made a realistic visual presentation for students. The cross section was 10"/48 or approximately 1/4" x 1/4". Thus, the area scale was 48x48 = 2304.

The materials used for the bridge were balsa or bass wood available from hobby stores which resulted in a material strength scale reduction of approximately 4. For this combination, the approximate scale, using the method of combining factors described in this chapter, would be 1/10,000 based on the area reduction and strength reduction. The final results are shown in the following, workbook section.

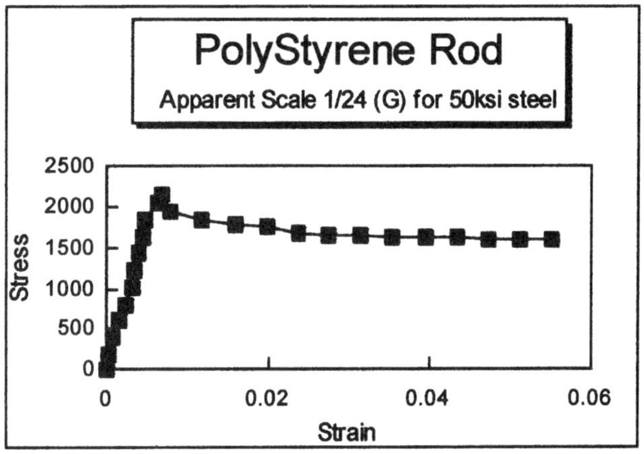

Figure 9.2 - Steel Scale Material: PolyStyrene

9.2 Strengths of modeling materials

Modeling materials can be developed and used in failure models. The stiffness in the materials can be adjusted. Since displacements in real structures are very small, using the same materials would make displacements in models too small to measure. For this reason, in structural models the stiffness is generally decreased to increase the models displacement. These models are referred to as strain enhanced models. It is common in these structural models to scale the stiffness with the scale.

The stress- strain plot for mild steel (Strength =36000 psi) was shown in Chapter 7. As an example of modeling materials, a plot of PolyStyrene is shown in Figure 9.2. Note that the strength of PolyStyrene (the stress at failure) is about 2100 psi. As some modern steels have strengths of 50,000 psi, this material would make a reasonable approximation for failure modeling of steel members in (G) scale. Smaller scales would need to seek a weaker material.

For reference, some typical construction materials, and possible modeling materials are shown in Table 9.2. The table shows typical construction materials and corresponding material properties on the left side of the table . Possible modeling materials and values found with lab testing are shown in the right side of the table. Note that the material properties were generally found with single tests and may not be considered typical values for these materials, but are included for comparison purposes. The approximate scale value is included in the parentheses in the table.

Table 9.3 was developed for modelers that may want to simulate failures at scale with continuous materials, such as concrete. The modeling material is based on Plaster of Paris combined with varying amounts of water and sand to vary the consistency and strength properties. The three mix materials are mixed on a volume ratio. So, for example, if a small amount of material was desired, a cup might be used as the measuring device. The mix on line 1 of the table would be made with 2 cups of plaster, 2 cups of sand, and 1 cup of water. If a larger amount of material is desired, a small bucket may be chosen as the measuring device with 2 buckets of plaster, 2 buckets of sand, and 1 bucket of water. The tested properties and approximate scales are tabulated. Figure 9.3 summarizes the data graphically and allows sand to be varied in the mix volume ratios to achieve strength values which may be desired. Two different plaster to water ratios are given to provide different consistencies of the final mix.

Table 9.2 Materials used in structural members

Construction Materials			**Modeling Materials**			
Material:		Typical values	Tested values and scale			
Rolled/Mild Steel			Acrylic	Buturite	ABS Plastic	Polystyrene
	E	30,000,000 psi	500,000	480,000	150,000	36,000
			(Approximate scale)			
			(1:64)	(1:64)	(1:200)	(1:800)
	Stress at Failure	36,000	9,200	6,400	3,000	2,000
			(Approximate scale)			
			(1:4)	(1:6)	(1:12)	(1:24)
	Strain at Failure	0.1-0.5	0.03	0.024	0.09	0.8
Woods						
Oak			Balsa	Bass		
	E	400,000	130,000	100,000		
			(Approximate scale)			
			(1/3)	(1/4)		
	Stress at Failure	8,000	4,000	1,580		
			(Approximate scale)			
			(1/2)	(1/5)		
	Strain at Failure	0.01	0.04	0.01		
Concrete			Plaster	Clay	Sugar	Ivory Soap
	E	1,500,000psi	14,200	1,200	21,500	3,000
			(Approximate scale)			
			(1:100)	(1:300)	(1:70)	(1:500)
	Stress at Failure	3,000 psi	800-1000	600	300	30
			(Approximate scale)			
			(1:3)	(1:5)	(1:10)	(1:100)
	Strain at Failure	0.0015	0.06	0.2	0.017	0.011

Table 9.3
PlasterCrete "Concrete" for Scale Model Mixes

Plaster	Sand	Water	Consistency	Strength	Stiffness	Strain	Comments
Ratio by Volume				Psi	Psi		
2	2	1	Dry	225	27500	0.003	
2	2	1.5	Milk Shake	150	35000	0.002	G 1/24
2	3	1.5	Dry	65	2500	0.003	
2	2	2					
2	3	2	Cake Batter	115	40000	0.003	G 1/24
2	4	2	Stiff liquid	70	15000	0.005	S 1/64
2	5	2	Cake Icing	15	1500	0.005	N 1/160

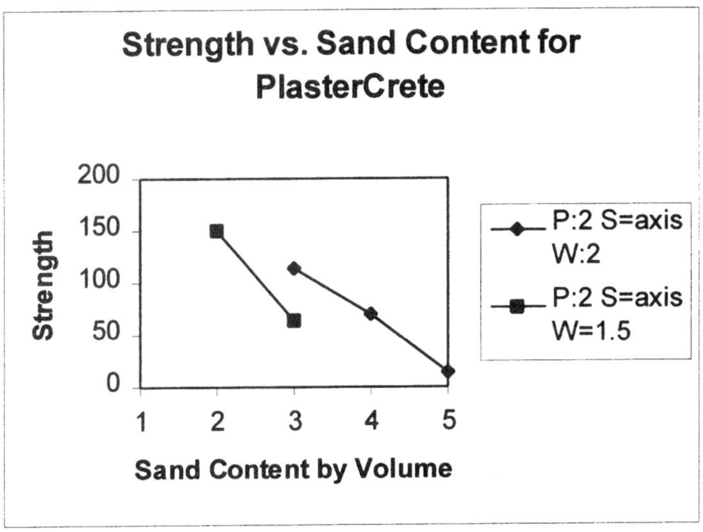

Figure 9.3 Failure Modeling Plastercrete

9.3 Time scaling

The similitude relations are not as intuitive for problems which require timed input such as vibrations and earthquake loadings. In general, these loading conditions require that time be speeded up in the model by a factor which is the square root of the model scale:

Model Time = Full scale time * $(\text{Scale})^{1/2}$

For example, a 24 second earthquake record would need to be placed on a 1/16 scale model in 6 seconds (the square root of 1/16 is1/ 4; 24* 1/4 = 6 seconds).

Figure 9.4 Student's Project Model ready for failure testing (see Workbook for Final and Fun results)

Chapter 10 - Summary

Figure 10.1 Old and New - Railroad Truss circa 1880 foreground
University of Colorado, Denver background

There are five major principles of Structural Engineering covered in the previous chapters. A brief summary of these principles follows below in sections 10.1-10.5. These principles are used to design many of the places people live, work, eat, go to school, travel, and have fun.

10.1 Spring Law: $Ky=P$

The Spring Law relates the displacement of a structure, y, to the loading, P. This relationship is most important in the analysis of structures like trusses.

10.2 Normal Stress: $\sigma = P/A$

The equation for normal stress calculates stress (or pressure) on a unit area of material. This relationship is most important to determine if forces will cause a material to fail. The structural engineer designs structures which will not fail.

10.3 Normal Strain: $\epsilon = y/L$

The equation for normal strain uses displacement within a member and relates this displacement to a length of material. Like stress, this relationship is important to determine if materials will fail.

10.4 Hooke's Law $\sigma = E\epsilon$

Hooke's Law relates stress to strain through the constant, E - Young's Modulus. Note that this is the equation for a straight line; Hooke's Law assumes materials are linear in nature. Values for Young's Modulus are tabulated in references used by engineers for analysis and design.

10.5 Stress in bending $\sigma = Mc/I$

The equation for stress caused in a beam when bending. These stresses are the same type as normal stresses. Along with normal stresses, these values are used to determine if a material in a structure will fail.

10.6 Trusses

The earliest trusses were based on principles from triangles and simple geometry. Simple mathematics can be used to sum forces to study the structures using components of forces in set directions. That is, the method of vectors, known from physics, is used. To assure that forces were only normal (along) members, pins were used in the actual construction of the early trusses. In this book a few examples of the early trusses built for the railroads were used as examples.

10.7 Modeling

For the modeler, geometric models can be made by shrinking dimensions of a real sized structures. These models are fun and useful in hobbies and films to make scenes that look like real life. Models that fail like real structures can also be made. These require the model and the real life structure to behave the same way on the material level. To make this happen the principle of Similitude, make similar, is used. In this book several methods to accomplish similitude of models are suggested. A new method, fractional similitude uses the different scale relationships from each material or component and combines these into one single number. This allows easy comparison with materials readily available.

10.8 Material Properties

Materials act in certain ways. Usually materials deflect under smaller loads in a uniform way. This proportional relationship of increase of stress to strain in structures is known as linear. Stress and strains which are linear plot as a line in the stress-strain graph.

Structures under small loads generally load and unload with the same strain; referred to as elastic.

Steel, when loaded to higher loads, yields, or stretches, in a manner which is no longer linear, but nonlinear. The ability to stretch with small load increases is the property of plasticity. While steel is plastic, it can stretch to large deflections before it fails.

Materials that break with small deflections, small strains, are referred to as brittle materials.

The stress at which a material fails is called the strength. It is related to the load direction; compressive strength and tensile strength as materials may be different in different load directions. As an example, concrete is approximately one-tenth as strong in tension as it is in compression.

10.9 Additional study - The Workbook

For those interested in studying these concepts in more detail and getting some "hands on" learning, the latter part of this book has some fun exercises.

Engineering can be fun.
It is all around us everyday.

References

—, Encyclopedia of Science & Technology 7th ed., McGraw-Hill, Inc., New York, 1992.

—, Knowledge in Depth, The New Encyclopedia Brittanica, Encyclopedia Brittanica, Inc, Chicago, 1976.

American Institute of Steel Construction, Manual of Steel Construction, 7th Ed., AmericanInstitute of Steel Construction, Inc.,101 Park Avenue, New York, NY 10017, 1970.

Beer, F.P. and E.R Johnston, Vector Mechanics for Engineers 5th ed., McGraw-Hill Book Company, New York, 1988.

Golenpaul, Dan (editor), Information Please Alamanac, 1965, Simon and Schuster.

Gustafson, David and Peter Frisk, Plane Trigonometry, Brookes/Cole Publishing Company, Pacific Grove CA. 93950.

Harris, David W. "Indeterminate Boundary Conditions: A Quadratic Programming Approach, A Thesis submitted to the Civil Engineering Department of the Texas A&M University, December, 1974.

Historic American Buildings Survey/ Historic American Record, Library of Congress, On-Line Historic Data, and Reproducible Archives.

Koerte, Arnold, Two Railway bridges of an Era, Firth of Forth and Firth of Fay, Birkhausen Vebley, Boston, 1991.

Mallery, Paul, Bridge and Trestle Handbook 4th ed., Carstens Publications, Inc., Newton, new Jersey, 1992.

McCormick, J.C., Structural Analysis, International Textbook Company, Scranton, PA, 1967.

Merdinger, C.J., "Civil Engineering Through the Ages," The Military Engineer, 1952.

National Building Code, Engineering and Safety Department, 85 John Street, New York, NY, 1967

Plastruct, Universal Model Parts Catalog, Plastruct, 1020 S. Wallace Place, City of Industry, CA 91748., Vol 7, 1998.

Reed, Robert C. Train Wrecks - A Pictorial History of Accidents on the Main Line, A Schiffer History Book, 77 Lower Valley Road, Atglen, PA 19310, 1996.

Sharma, V., J. Choros, and A.J. Reinschmidt, "Static and Dynamic Testing of a Through-Truss Bridge," Report No. R-806, Association of American Railroads, AAR Technical Center, Chicago, Illinois, 1992.

- "The failure of an Old Railway Bridge in France", Engineering News, April 9, 1908.

Van Zandt, J.G., "Outline History of Railway Bridge Building in the U.S.," Railway Engineering and Maintenance of Way, Oct. 1912, pp. 339-459.

Waddell, J.A.L, Bridge Engineering, 2 Vols., John Wiley and Sons, New York, 1916, p 477

Contributors to Text

American Institute of Steel Construction, Inc.
101 Park Avenue
New York, NY 10017

Transportation Technology Center, Inc
A subsidiary of the Association of American Railroads
55500 DOT Road
Pueblo, Colorado 81001-0130

Colorado Historical Society
Colorado History Museum
1300 Broadway
Denver, CO 80203-2137

Model Railroader Magazine
1-800-533-6644

Ontario & Western Historical Railway Society, Inc.
Box 713
Middletown, NY 10940

Plastruct
1020 S. Wallace Place
City of Industry, CA 91748

The John Maxwell Collection
1055 Jessica
Fernley, NV 89408

The Library of Congress
Washington, D.C.

University of Florida

Workbook: Fun For Students - Projects and Problems

Fun reading for those interested:

This section of the book is intended for students, or anyone, who wants to study the concepts presented in this book in more depth. There are the following fun things to do:

1. Problems for practice. It's fun for students to see if they can work some of the problems from the book.

2. Simple experiments which can be done at home, at school, or in clubs and groups.

3. Advanced projects for an in depth look at some of the topics from the chapters.

Answers to selected problems follow the workshop material.

Truss Fun

Chapter 1 List exercise:

Make a list of trusses you see around your home:

1. ______________________________
2. ______________________________
3. ______________________________
4. ______________________________
5. ______________________________
6. ______________________________
7. ______________________________
8. ______________________________
9. ______________________________
10. ______________________________

Make a list of trusses you find in books, movies, magazines

1. ______________________________
2. ______________________________
3. ______________________________
4. ______________________________
5. ______________________________
6. ______________________________
7. ______________________________
8. ______________________________
9. ______________________________
10. ______________________________

My favorite truss is:

____

Most unusual truss:

____

Chapter 2 Drawing some trusses:

Connect the dots with a:

```
     O    O    O    O

  O    O    O    O    O
  ^                   ^
```

Warren Truss

```
       O    O    O

  O    O    O    O    O
  ^                   ^
```

Pratt truss

```
       O    O    O

  O    O    O    O    O
  ^                   ^
```

Howe truss

Would these be deck or through trusses?

Chapter 3 Project: Quick Trusses

The "true" Warren Truss is made of equilateral triangles (equal length sides). This makes an ideal quick project to study the triangle building block. Toothpicks, pointed on both ends, can be used as the members of the truss. Anything that will hold the toothpicks can be used as nodes. In the end, a truss is produced as shown below. Note that there are two sides, identical trusses, which can also be connected by members (toothpicks) to make a 3 dimensional truss which can stand up. With very little work, a fun structure can be made.

A Carrot Warren

Cut carrot pieces of carrot about 3/8" thick (and round). Use these as your joints. Really quick and easy. Let your kids be Godzilla and eat the bridge, adult supervision with the toothpicks, please. Great for finicky eaters.

The Carrot Truss over a Lettuce River

Styrofoam Warren

Use styrofoam for permanent joints if you want to keep the bridge.

Some games and contests

Try to build the longest span before failing.

Most load: Make a truss which spans your kitchen sink, or a large pot. Put a small pot on the top. Carefully add water until it fails. Use a table to determine the weight, see constants next page.

Volume	Column A Weighs	Column B How many?	Subtotal weight (AxB)
1 cup	0.5 Pounds		
1 quart	2.0 Pounds		
1 gallon	8.1 Pounds		
Total Weight	XXXXXXXXXXXXXX XXXXXXXXXXXXXX XXXXXXXXXXXXXX	XXXXXXXXXXXXXX XXXXXXXXXXXXXX XXXXXXXXXXXXXX	(Sum of above rows)

Table for converting volume of water to weight

Additional Information: Metric Units

In the preceding, chapters English units - pounds and feet were used. The main reason for this use was that the historical materials used in the book were in these units. Today metric units in the International System of Units, SI, are used throughout the world and on many projects. The base units in this system are the meter and the kilogram. The following table is offered to aid in converting English units from this text to standard SI units .

To convert	To:	Multiply by
Feet	Meters	.3048006
Inches	Meters	.0254
Square foot	Square meters	.09290341
Square inches	Square meters	.0006451600
Pound - force	Newton	4.448222
Kip - force	Newton	4448.222
Psi	KG per Square meter	703.0696
Psi	Pascal	6894.757
Kip-Ft	Newton-meter	1355.5818
Pound-ft	Newton-meter	1.355818

Ref: US Bureau of Reclamation. Metric Manual, 1978

Chapter 4 - Historical Research

Find information about a real truss in the library, or on the Internet. Write a short summary about the truss.

Truss Name: ____

Location: ____

(Picture)

Outline of writeup

I. History
- Date
- Location
- Type
- Designer/Builder
- Initial use/purpose

II. How used today?
- Same location?
- How much use? Number of cars, Converted to Historic Pedestrian bridge etc.

Historical Information from the Internet

There are many sources of information on the Internet about trusses.

Library of Congress

The Library of Congress is one of the largest sources of information available in the United States. Much of the information is free, can be used in reports, and is at your fingertips. The site is www.loc.gov . The site changes and has improvements over time. In general, you can use a search using multiple words in the entire collection, or go directly to the Historical Engineering Records and find lots of information on bridges.

Advanced project

Write an account of a bridge similar to those in Chapter 4. Research and find pictures and drawings of the truss.

Chapter 5 Practice Problems

1. You have an action figure which is 6" tall. What scale is the action figure compared to your height?

2. The scale for HO trains is 1/87. If a typical model locomotive is 6" long, how big would the real locomotive be?

3. A sugar cube is approximately 1/2" on each side. If sugar cubes were used to build a scale model block wall which in full scale used 6" x 6" cinder blocks, what would be the approximate scale?

4. How much should an exact model of a Cooper E40 locomotive weigh in the following scales

E40 Locomotive total wheel weight = __________

G scale	(1/)	=
O	(1/)	=
HO	(1/)	=
N	(1/)	=

Why is this different from models you can buy?

5. Using Figure 4.26 (Lyons Brook Bridge) what G scale modeling product would best resemble the cross braces in the truss? What other section might be used in a model?

Advanced Project

Build a model based on drawings from a real bridge.

Sources of Additional Information

Organizations:

National Model Railroad Association
4121 Cromwell Rd.
Chattanooga, TN 37421
Home Page: www.nmra.com

(G scale)
Bay Area Garden Railway Society
205 Crystal Springs Center, Suite 40
San Mateo, CA 94402
Home Page: www.bagrs.org

(HO and N)
The Model Railroad Club, Inc.
Jefferson Ave., PO Box 1146
Union, NJ 07083
Home Page: www.tmrci.com

Suppliers

Model Railroader Magazine
800-533-6644
Email: Mrmag@mrmag.com

Plastruct
1020 S. Wallace
City of Industry
CA 91748

Chapter 6 Analysis of Trusses

For the structure below:

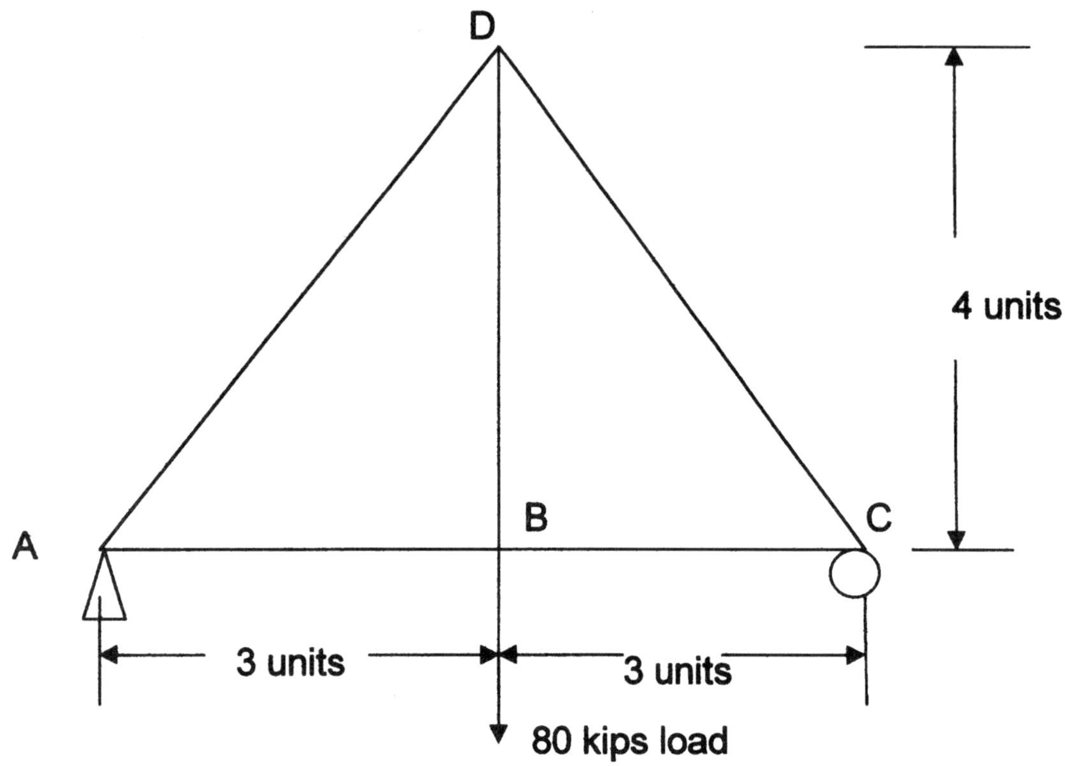

1. Find the load in each member by the method of pins
2. Check the upper chord load by using the method of sections
3. Check all members using program TF2DFE (see order blank on last page of book for available disk).

Spring Problems:

1. A ballpoint pen takes ½ pound of force to push the spring in the pen and make the point travel 1/8 inch . What is the spring constant of the pen?

2. A railroad car has 4 springs on each wheel system (there are 4 wheel systems) to carry the weight of the car. A railroad car holds about 125,000 pounds. If a railroad spring compresses about 1 inch, about what spring constant would be needed for the springs?

Advanced Project:
Analyze the 48 foot standard Howe Truss in Chapters 4 and 5. Use Cooper E40 loading. Place locomotive wheel loads on the bridge up to the distance of 48 feet. If wheel loads do not land on a node, proportion the loads to the nodes. For example, if a 40 kip load falls three feet from one node and 5 feet from the next, put 40*(5/8) = 25 kips on the closest node and 40*(3/8) = 15 kips on the farthest node.

Additional information: Mathematical Aspects of the Finite Element Method

In algebra, equations often have several variables. For example:

$$2x_1 + x_2 + x_3 = 16$$
$$-x_1 + x_2 - x_3 = -6$$
$$-4x_1 + x_2 + x_3 = 4$$

These are called simultaneous equations, multiple equations which contain the same variables and must be solved at the same time to obtain an answer. In algebra, these equations may be solved by substitution, solving for a variable in terms of the other variables and using this relation to substitute into a subsequent equation (example to follow).

There is another type of algebra referred to as linear algebra. Some of the methods in linear algebra simply use standard algebra techniques and a shorthand to write equations and keep track of operations, called a matrix. The above problem would be rewritten as:

$$\begin{vmatrix} 2 & 1 & 1 \\ -1 & 1 & -1 \\ -4 & 1 & 1 \end{vmatrix} \begin{matrix} x_1 \\ x_2 \\ x_3 \end{matrix} = \begin{vmatrix} 16 \\ -6 \\ 4 \end{vmatrix}$$

where each row of numbers is the same as above and each column of numbers are the values for a given x - x_1, x_2, and x_3 respectively. In linear algebra, there is a method to solve for each of the variables known as Guass Elimination. This is a step by step way to produce a matrix with 1's on the diagonal of the matrix, zero values below the diagonal, and transformed numbers above the diagonal. Once, this form is achieved, all values may be readily found using backsubstitution of values which are known starting with the last row. To summarize the steps:

1. Form a one on the diagonal by dividing by the value on the diagonal
2. Eliminate the value in the column below the diagonal, call the row with the 1 in the diagonal row d, by multiplying row d by the value (value c) in a given row, row w, under the diagonal column times -1. Add row d of the working row , row w.
3. Multiply row d by value -1/c to reform a 1 in the diagonal.
4. For row w, return to step 1

Step 2 sounds really complicated, but it is easy to see in an example. Below is an example which shows how to use this method and shows that it is the same as algebraic substitution with another name.

Solve the example using:

Algebraic Substitution	Guass Elimination
Divide Equation 1 by 2 $x_1 + x_2/2 + x_3/2 = 8$ (All other equations are the same)	Matrix starts as: $\left\vert\begin{array}{ccc}2 & 1 & 1\\ -1 & 1 & -1\\ -4 & 1 & 1\end{array}\right\vert = \left\vert\begin{array}{c}16\\ -6\\ 4\end{array}\right\vert$ Form a 1 on the diagonal, row 1 Divide row 1 by 2 $\left\vert\begin{array}{ccc}1 & ½ & ½\\ -1 & 1 & -1\\ -4 & 1 & 1\end{array}\right\vert = \left\vert\begin{array}{c}8\\ -6\\ 4\end{array}\right\vert$
Isolate x_1 in terms of other variables $x_1 = 8 - x_2/2 - x_3/2$ Substitute into equation 2 $-(8 - x_2/2 - x_3/2) + x_2 - x_3 = -6$ Or $3/2\ x_2 - 1/2\ x_3 = 2$ (Equation 3 the same)	Rule 2: Multiply row 1 by -1*(-1), add to row 2 Rule 3: Reform 1 in diagonal of row 1 $\left\vert\begin{array}{ccc}1 & 1/2 & 1/2\\ 0 & 3/2 & 1/2\\ -4 & 1 & 1\end{array}\right\vert = \left\vert\begin{array}{c}8\\ 2\\ 4\end{array}\right\vert$
For convenience multiply equation above by 2/3 $x_2 - 1/3\ x_3 = 4/3$	Rule 1 for row 2: divide to form 1 in diagonal $\left\vert\begin{array}{ccc}1 & ½ & ½\\ 0 & 1 & -1/3\\ -4 & 1 & 1\end{array}\right\vert = \left\vert\begin{array}{c}8\\ 4/3\\ 1\end{array}\right\vert$
For equation 3 substitute x_1 $-4(8 - x_2/2 - x_3/2) + x_2 + x_3 = 4$ Or $3x_2 + 3x_3 = 36$	Rule 2: Multiply row 1 by -4*(-1), add to row 3 Rule 3: Reform 1 in diagonal in row 1 $\left\vert\begin{array}{ccc}1 & ½ & ½\\ 0 & 1 & -1/3\\ 0 & 3 & 3\end{array}\right\vert = \left\vert\begin{array}{c}8\\ 4/3\\ 36\end{array}\right\vert$
For equation 3, substitute for x_2 $3(4/3 + 1/3\ x_3) + 3x_3 = 36$ Or $4\ x_3 = 32$ $x_3 = 8$	Rule 2: Multiply row 2 by -3, add to row 3 Rule 3: Reform 1 in diagonal in row 3 $\left\vert\begin{array}{ccc}1 & ½ & ½\\ 0 & 1 & -1/3\\ 0 & 0 & 4\end{array}\right\vert = \left\vert\begin{array}{c}8\\ 4/3\\ 32\end{array}\right\vert$ $\left\vert\begin{array}{ccc}1 & ½ & ½\\ 0 & 1 & -1/3\\ 0 & 0 & 1\end{array}\right\vert = \left\vert\begin{array}{c}8\\ 4/3\\ 8\end{array}\right\vert$

Solve for x_2 $x_2 = 4/3 + 1/3(8)$ $x_2 = 4$	**Back substitute for x_2** $\begin{vmatrix} 1 & ½ & ½ \\ 0 & 1 & 0 \\ 0 & 0 & 1 \end{vmatrix} = \begin{vmatrix} 8 \\ 4 \\ 8 \end{vmatrix}$
Solve for x_1 $x_1 = 8- 4/2 - 8/2$ $x_1 = 2$	**Back substitute for:** $\begin{vmatrix} 1 & 0 & 0 \\ 0 & 1 & 0 \\ 0 & 0 & 1 \end{vmatrix} = \begin{vmatrix} 2 \\ 4 \\ 8 \end{vmatrix}$

Simultaneous equations problems:

1. Solve the following:

$$x_1 + x_2 = 25$$
$$2x_1 - x_2 = 5$$

2. Solve the following:

$$2x_1 + x_2 + x_3 = 7$$
$$x_1 + x_2 + x_3 = 8$$
$$5x_1 + x_2 + 2_3 = 8$$

a. Using algebraic substitution
b. Using linear algebra.

Note that problems are easy to make, just assign values to x variables, write the left hand side of the equation, solve for the right hand side. The solution is easy to check, if you know the answer ahead of time, as each step can be checked by substituting in the known values and the solution should agree to the assigned values.

Finite Element Programming Projects: Projects to improve TF2DFE

Beginner Level:

1. Add LPRINTs to the program to produce a hard copy of results.
2. Write a subroutine to ask questions and guide user through input
3. Save an input session to a file.
4. Rewrite input to choose between data statements, a prepared file, or use subroutine to answer questions.
5. Make program loop and restart to solve multiple problems.

Intermediate Level

1. Plot the truss using the coordinates X,Y and the node numbers E1,E2
2. Make variables to save the maximum tension (+) and compression (-) by element and material group

Advanced Level

1. Add graphics to plot a picture of the boundary condition, and the loads with values, or
2. Add a load type of uniform load on members, calculate equivalent loads at the nodes and solve, or
3. Add the stress for members using the equation in Chapter 7 for normal stress = P/A
4. Plot maximum value of load on each member.

Chapter 7 Practice Problems

1. How many pounds per square inch are created by a 5 pound bag of sugar laying flat? (Estimate or measure the size)

2. A 6" (diameter) x 12"(height) cylinder is tested in compression in the laboratory. The cylinder breaks at a load of 140,000 pounds. What is the stress when the cylinder breaks? The same cylinder deflects 0.02" before it breaks. What is the strain when the cylinder breaks? What is the Young's Modulus for this material?

3. A ½ " diameter steel bar has a strength (the stress at which the bar will break) of 36,000 psi. How much force will it take to break the bar? If the Young's Modulus is 36,000,000 psi, according to Hooke's Law, at what strain will the bar break. If steel is ductile, why might the strain be greater?

4. My 1986 Ford pickup weighs 8600 pounds empty. The tires (remember there are 4 tires)form an area on the pavement of 9" x 9". How much stress does my truck make on the pavement? I can carry a maximum load of 6,000 lbs. How much total stress would this make on the pavement? My tire pressure would have to be 65 psi. Is this bigger or smaller than the stress? Why would it need to be?

5. A beam is in a simple span of 30 feet. A highway deck creates a uniform load of 100 pounds per foot on the beam. What is the moment at the midspan. If the beam is a rectangular section that is 1 foot wide and 2 feet high, what is the stress in the beam.

Chapter 7 - Testing Materials at Home

Materials are tested in laboratories using large specialized equipment. College engineering students and some high school students will be using testing equipment to fail and study various materials. Some simple experiments can be done in the home or in school labs to show the idea of testing. These same experiments can be used to demonstrate material properties using simple materials.

Testing strength and showing Plastic behavior

Steel is a typical material which is very strong and yields, that is - shows plasticity, or strains a visible amount before it breaks. To make a test at home, a material which acts like steel is clay. Take some modeling clay and make 3 small cubes about 1/2" on each side, as close to identical as possible. Place the squares between pieces of paper (to save a mess) and set a large pot on top. Measure the distance from the pot to the counter top. Slowly add water to the pot, say about a quart at a time. After each quart, measure the distance from the pot to the counter (Note if a quart makes a large deflection, make the cubes bigger, if after several quarts - nothing happens, make the cubes smaller). Write your measurments in the table on the following page.. The squeezing out of the clay is called flow, or plastic flow.

Testing strength and showing Brittle behavior

Concrete is a typical material which is mixed from a number of other products. Concrete fails in a brittle manner. At home, brittle type behavior may be seen using small sugar cubes. Measure the dimensions of a cube. (Note sugar cubes are amazingly strong - several lighter weight folks may be needed for this experiment). Place 1 sugar cube between two boards (pretty large boards) and stand on the top. The sugar cube should (hopefully) fail suddenly. Now try 3 cubes. (Note that three cubes can be spread out in a triangle which is more stable). Keep adding people until the cubes fail. Is there much difference in the height of the boards before the cubes fail? (Brittle materials fail with very little strain, which is very little deformation). Weigh everyone that helped fail the cubes, that total will be the failure load P. Use the area of all cubes and find the failure strength of the sugar as the stress at failure, or

Strength = Normal stress = P/A

Truss Fun

Plot your own stress- strain curve below based on your results.

Clay Cubes:
Dimension = ____________ Inch (or metric =) on all sides Area = area of cube x number of cubes= ___________

Volume: 1 (Quart, etc) = -------------- Lbs. (or metric)	Total Weight of Pot and this Volume of water	Total Deformation for this volume (Change from initial height of pot)	Stress = <u>Total Weight(Col.2)</u> Area	Strain = <u>Deformation (Col 3)</u> Initial Dimension
1.				
2.				
3.				
4.				
5.				
6.				
7.				
8.				
9.				
10.				
11.				
12				

Stress- Strain Plot of Clay Material

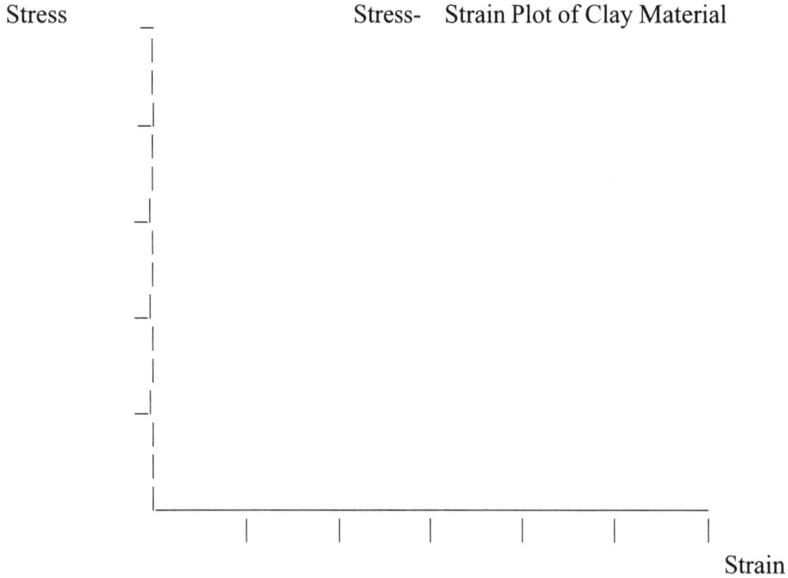

Chapter 8 Project: Testing Structures at Home

We can use a large kitchen pot as our test stand and we'll be able to measure many properties. We'll use water as our loading, so we'll need two large measuring cups.

Testing Connections

Place two rulers across the lip of the pot. Make test joints with 2 carrot pieces connected by a toothpick, like the carrot truss. Hang this arrangement from the two rulers into the pot. Use rubber bands to hang a measuring cup from the lower joint. Fill the cup in the pot with water until the toothpick pulls out of the joint material. Record this weight as the tension/pull-out load for the joint material.

Testing a Structure

Make a cube with toothpicks and joint materials (carrots or styrofoam). Place your pot on top of the mini-structure. Carefully fill the pot until something drastic happens. Hold pot gently as you fill. Something is going to happen suddenly -fill the pot slowly. What happened? What kind of failure is this?

Make a new cube with triangular cross braces. Test the new cube. Did this cube take more or less weight? Why?

Buckling

Find a tall and slender member for your truss, say a bamboo skewer for barbequing, or a wire, or a thin dowel rod. Try the same experiment as above with a tall rectangular tower. If the joints fail first due to moment, try to steady those gently with your hand. Produce a buckling failure in your tower members. Record the buckling load for that member.

Volume	Column A Weighs	Column B How many?	Subtotal weight (AxB)
1 cup	0.5 Pounds		
1 quart	2.0 Pounds		
1 gallon	8.1 Pounds		
Total Weight	XXXXXXXXXXXXXX XXXXXXXXXXXXXX XXXXXXXXXXXXXX	XXXXXXXXXXXXXX XXXXXXXXXXXXXX XXXXXXXXXXXXXX	(Sum of above rows)

Design principles:

What changes in design were needed to make the New Tay Bridge safe?
Why are standard methods needed for engineers?
In the case history of the bridge failure in France. Could the loss of lives been prevented? How?
Fun from the Net:
Visit the University of Florida "Civil Engineering Truss Bridge Laboratory": (Link from Educational CE Lab Activities). Build and test simple popsickle stick structures.
Balsa Wood Models to test: University of Syndey: Look for Bridge_truss_project
Toothpick bridges:
Yes Magazine, CA Newton's Apple look for bridges

Chapter 9 Materials modeling

What material acts most like concrete in G, HO and N scale?

Advanced project

Build the 48' Standard Truss to G scale. How long will the bridge be? Change member dimensions in both directions by the scale. What will the estimated final model scale be? Use water as load spaced to simulate 2 locomotives. Load to failure.

A Student Project

In the summer of 1999 a student project was completed as part of the Colorado School of Mines, Summer Minority Engineering Program, know as SUMMET. A student project based on the principles of Truss Fun was suggested and accomplished by one group of students. This project was conceived to incorporate many of the principles introduced in this book.

An historical truss, similar to Chapter 4, was selected. The scale of the drawing chosen was 1/4" = 1 foot, or a geometric scale of 1/48 (Chapter 5). This scale was used to build the selected model. In this manner, the blueprints provided a picture of the actual size of the model. The prints were used to size members and serve as a layout and cutting guide. Sizing the rectangular members to the drawing created an additional reduction in the area of the members. The length scale was 1/48, the area was reduced by 1/48th in each direction, or by 1/48 times 1/48.

The fractional similitude was estimated before building the model, using the principles from Chapter 9. Members for the truss were scaled by using a 1/4" by 1/4" basswood piece to approximate 10" by 10" and 12" by 12" oak from the actual truss. The area reduction becomes approximately 1/48 by 1/48 , or 1/2304. The stiffness of the basswood is estimated as 1/4 of oak from Table 9.1. The final scale for failure becomes AE/L * Y = P or 1/2304* 1/4 y = P. The final load would scale to approximately 1/10,000. Following the construction and failure of the model, comparisons were possible as discussed following the students work.

The project took about 1 week (7 days) to complete. The student's report (with minor changes to [hopefully] aid in clarity of the discussion) is contained in the following text. The students were asked to name themselves and act as an engineering company. As shown in their report, they referred to themselves as Lost & CO. This report discusses the railroad bridge in a real context. The high school students took the project seriously, for which I thank them, but inclusion of the full report text is for illustrative and educational purposes and is not intended in anyway to be a commentary or draw engineering conclusions with regard to railroad bridges in use. This project is typical of efforts possible based on the concepts of this book.

STRUCTURAL PROJECT LOST & CO.

ASSOCIATES
Erin Hankle
Ella Schmidt Jessica Gonzalez
Amoret Lambrecht Bridget Trostel James Sanchez
Carlos Medina

Truss Fun

Executive Summary

Bridges are a problem. The dependency of old bridges is being questioned in order to determine if they are in need of renovation. For example, the Howe truss design was common in the late 1800's, and posed numerous threats. These threats varied from loss of lives, damaged goods, and devastation to designer credibility. The aging infrastructure's largest problem today is dealing with modern loads. The deterioration of old bridges leads to our motivation of discovering their maximum stress capacity.

The approaches LOST & CO applied to this situation were based on a physical and a computer model. The physical model of the Howe truss was designed on a 1/48" scale with slightly varied materials. The small scale limited our material possibilities and required us to change a few design methods, such as replacing metal braces with Epoxy and adding one foot long horizontal pieces of wood adjacent to the spliced joints. The computer model was directly related to the 1890 bridge design and demonstrated results similar to our physical model.

The main findings of our research project concluded that the computer and physical models were related. When each model was tested, they both displayed a similar failure point and displacement. Both models displayed a factor of safety of approximately 2, failure point at 185 lbs., and a maximum displacement before failure of 0.7 inches. Though the results showed the bridge to be strong enough for a typical 1890 load, the comparison demonstrated that the bridge is incapable of holding modern train cars.

We recommend the use of additional reinforcement and supporting beams if this bridge is going to be used under modern stresses. A vertical brace must be placed below the bridge at its center to stabilize the structure. Strengthening material should be arranged along the bottom of the bridge to prevent cracking and unnecessary bending that would weaken the truss. If these changes are not made, the bridge will ultimately fail and cause much devastation and possible loss of human lives.

Workbook: Fun for Students - Projects and Problems

I. Introduction

In 1840 William Howe of Spencer, Massachusetts patented a new type of truss system. A farmer turned inventor, he adapted the Howe truss from an earlier truss design. He inserted vertical iron rods into the plan to carry the tension loads and wooden diagonals to carry the compression loads. This allowed for much greater distances to be bridged. With this new system the distance spanned could be increased up to 200 feet by simply adding more diagonals and vertical rods. Some favored the truss's easy construction and low cost. The Denver Rio Grande Railroad used the Howe truss regularly on their railways. But the Howe truss was met with friction from others. "...the Howe truss was probably responsible for as many railroad deaths as any other single cause. Although the engineering and construction of these truss bridges was probably not always the cause of their failure, there is no doubt that they were particularly vulnerable to floods, fires, and decay. They required constant inspection and repair. Also a number of bridge collapses can be traced to derailed trains striking the timber trusses and weakening the bridge enough to cause it to fail" (Reed, 85). Was it massive media sensationalism or the honest truth? Through our experimentation we plan to put an end to the speculation. We will test a model bridge to failure and compare the loading to historical and modern day loads to decide whether the attack on the Howe truss is justified. Our experimentation may also be used to evaluate the aging erials or lives due to the collapse of bridges and other structures.infrastructure of this country and suggest courses of action for repairs and rebuilding. Applying our results will prevent the loss of materials or lives due to the collapse of bridges and other structures.

II. Objective

We, LOST & Co., were approached with a proposition to construct a model bridge from a set of blueprints created in 1890. Though the blueprints are still intact, the existence of the designer and actual bridge still remains a mystery. The problem we faced was to determine whether, if the bridge was built and still stands, it could maintain its stability while being subjected to the pressure of a modern day train load. If the bridge failed at a modern load, we would next determine if it was safe at the time of construction and thereafter.

The project was necessary to check the safety and security of the bridge. Many people would depend on this bridge if it existed on a railroad line. The trains running across the bridge would carry various supplies and other goods needed to maintain a good quality of life in the towns and communities surrounding the railroad. While it is still unclear whether the bridge stands today, the need for these tests are still apparent. The factor of safety must be established because if the bridge was ever constructed, we must know if it would stand up to the pressure of the trains crossing. All parties associated with the bridge would share benefits of this process. The railroad company will be saved the hazard of a possible derailment and costly crash. The goods and services on the trains will reach their destinations intact.

III. Design Method

We used a blueprint from 1890 of a 202' Howe Truss drawn to 1/48" scale as the basis for our model test bridge. The test model was also built to 1/48" scale of the actual bridge, thus shrinking the area to 1/2304. The bridge was constructed with 1/4" balsa wood, 1/8" bass wood, 2 ½" screws, bolts and nuts, 3000 p.s.i.Epoxy, Mylar tape, and copper wire. The final measurements of the bridge were roughly 5' long by 5" high by 3" wide. The joints were made with ½" bass wood dowels and wood glue.

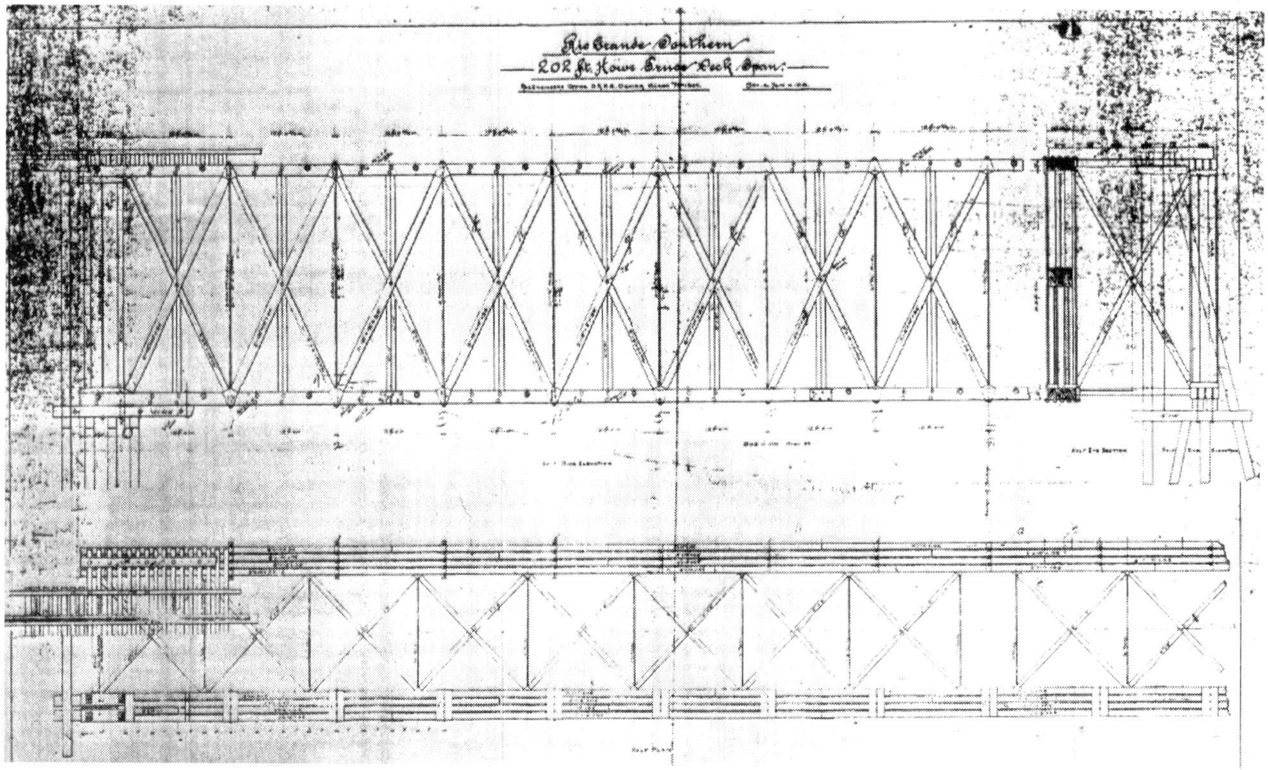

Project Figure: A Howe 202' Truss Blueprint
Reprinted with permssion John Maxwell Collection, 1055 Jessica, Fernley, NV 89408

In addition to the design methods, the joints were reinforced with inch long bass wood dowels. The dowels were spliced into the ends of the balsa wood, one-half inch into each side, joining the two pieces. Also, a one-foot piece of balsa wood was layered adjacent to the joints for extra support. The joints were offset to reduce possible vulnerable places and failure at these key positions.

The metal braces that were used in the blueprints to strengthen the connections were not available to us in the size desired. Epoxy, at a strength of 3,000psi., was utilized to replace these braces. It increased the bridge's relative strength. Steel rods were not available for employment, so six-inch steel bolts and washers were added to provide vertical support.

Solid squares of balsa wood became substitutions for the trusses that were positioned in the center for lateral support. Lateral support was not an issue we wished to test in this experiment. Only the bending strength results of the structure were desired by the company who contracted the experiment. The blocks were only present to prevent the bridge from tipping.

IV. Construction Method

The construction of the bridge was complicated due to its significantly small scale. To make this task easier, we chose materials matching the exact dimensions and scale of the blueprint. We began the construction by pre-drilling holes for all of our screws and cutting the horizontal deck pieces roughly to length. Upon completion of this stage, we started on the problem of connecting the deck pieces end to end. The group decided to drill long holes into the ends and form splices between the components. We ended the first day while waiting for the glue to set.

We began the next day reinforcing the joints between the splices with both Mylar and copper tape. The team then started installing the screws in the deck pieces. A few miscalculations in the design led to the exhausting process of backing out all of the screws halfway. This error in judgement and its remedy occupied most of the afternoon, but later in the day we were able to lay out the cross brace system and have it ready to install the next day.

The following day we worked on gluing the cross members in place. We glued a few braces in place at a time so we could hold each until it was set in the glue. This guaranteed good contact between the braces. Then we split into two teams to continue with the construction. The first team proceeded by gluing the second set of cross braces into the truss. When the second beam was put in place and the screws installed, we realized that the holes did not line up with several of the screws. We re-drilled these holes to remedy the situation. The second team began duplicating the opposite half of the truss system. A problem became apparent early on in the process. None of the screw holes in the beams were lining up. This issue was more difficult to repair. The group was forced to use a razor blade and separate the cross members from the glue and reposition the horizontal beams. After resolving these difficulties, the two teams ended the day at a similar stage for each truss section.

When we began work again, both teams proceeded attaching the outer horizontal deck components. This along with inserting and positioning screws took a majority of the day. Having completed this, both construction crews inserted steel bolts to take the place of the cables and rods that were present in the blueprint. These were necessary to provide vertical tension support for the bridge.

Project Truss - The finished bridge ready for testing

Both sides being complete, the teams decided to reinforce the joints and the connection site. The placement of another piece of wood next to each beam helped strengthen the bridge at one of its most critical points. The wood was cut and glued into place with no dilemmas. We created this reinforcement because the blueprint displayed the use of steel plates. Not having the plates to endure some of the pressure, we improvised. Next, to conserve time, the group came to a consensus and voted to use five solid pieces of balsa wood instead of trusses. Not needing to test the lateral strength of the bridge and only the bending, this was only to keep the structure from tipping to one side or the other. The two components were brought together and pasted into place with the epoxy. Then, wrapping the crossbar joints with copper wire, the bridge was set aside to cure. We considered the bridge to be complete and a work of art.

V. Test Method

Our group decided we were going to test the load of our bridge by using buckets of water. The buckets were weighed and marked at ten-pound increments for easy measuring and testing. Our initial thought was that the edges of the buckets would cut into the wood and weaken the structure. We planned to use bricks as a first layer to avoid this problem, but on the day of testing, we realized that the washers would serve the same purpose. We loaded the buckets in a pyramid form, with only light support from our fingers to steady them. This support provided the stability to test only the bending strength and not the lateral strength. The four buckets were loaded from the middle to the outside adding ten pounds of water in each rotation. When the base buckets were filled to the thirty-five pound mark, we stacked three buckets in another layer on top of them. They were filled to twenty pounds each before the initial failure. Upon failure, the bridge collapsed and the buckets tipped over spilling the water.

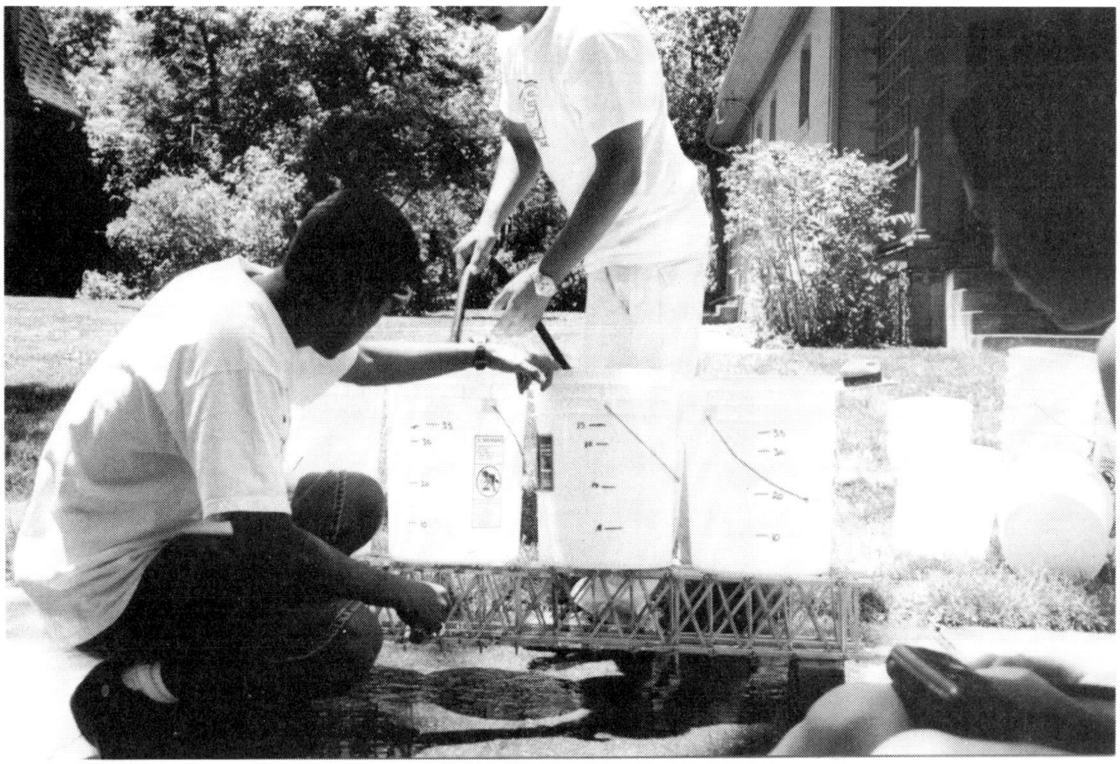

Project Figure(added) - Bridge being tested with buckets of water

VI. Results

We began the testing at 1pm on June 26, 1999. We set the bridge up on top of wooden stands. We had pre-measured buckets to fill with water as the weight to be applied. We first set four empty buckets on top of the bridge. We added 10lbs.of water to the middle buckets first and then to the outside. After adding the first 40lbs. we took the displacement measurement. We started out at 3.2" and after the 40 lbs. it was at a displacement of .1". We then added the next 40 lbs. of water into the buckets. The displacement barely went down to .2". We kept adding weight. The displacement stayed at .2" for the next 90lbs. We finally got to 185 lbs. when the bridge began to fail. It was failing at the bottom support beam. Here it had a displacement of .7". We added more weight to totally have the bridge collapse. It did not collapse until we got the bridge at 200 lbs.

Results:

Displacement(in.)	Weight(lbs.)
0	0
0.1	40
0.2	80
0.2	120
0.2	140
0.2	170
0.7	200

Project Figure (Substitute for another picture) - Detail of failed bridge

VI. Conclusion and Recommendations

The media slander on the Howe truss bridge style is not justifiable. During the late 1880's and early 1890's the maximum train load was well below the truss systems limit of 445,000 lbs. As for the modern loads averaging about an **E72** the 1890 truss wouldn't stand a chance according to our testing.

If more information was known, such as the existence, whereabouts, condition, types of support, and wear, the conclusions we have reached could vary dramatically. The design consisted of three supports. In the field and in the lab the third support could make all the difference. Again in the model the joints were not exactly similar to these on the blueprint. The minor changes were made because of the time limitations given to us and the amount of time it would have taken to place so much detail onto the model.

Our recommendations for this experiment as well as building and/or maintaining this truss system in the field are as follows.

Modeling

1. Do not use balsa wood in the model. This will weaken the structure and lead to certain failure.
2. Do not pre-drill holes in all load-bearing beams. It becomes a problem to align these holes in the final assembly.
3. Utilize strong Epoxy to replace metal braces. Be sure to work in a well-ventilated area.
4. Account for all materials beforehand. Inventory supplies and buy only what is needed.
5. Have spare materials ready in case of accidental breakage.

Field Construction

1. Add multiple vertical supports beneath the bridge to increase the strength.
2. Increase the support on the ends of the bridge.
3. Reinforce joints with another beam spanning across the two existing beams to add support and strength.

Author's Notes - Following Project Completion:

Report Reference: Reed, Robert C., Train Wrecks - A Pictorial History of Accidents on the Main Line, Schiffer Publishing Ltd., 77 LowerValley Road, Atglen, PA 19310

Following the model failure, additional work was completed to study the final similitude and results of the model. The Finite Element Computer Program in Chapter 6 was used. Several runs were made with the computer code, varying the Young's Modulus, E, each time until the displacements in the model were matched. A value of E of 80,000 psi was found using this approach. The ratio of this parameter using Table 9.1 values for Oak was thus 1/5 as opposed to the assumed 1/4. The final fractional value is 11,520.

The model failure load of 185 pounds can be converted to field loads using the fractional similitude, or 185*11520 = 2,133,050 pounds in the field. A 1880 train on the span would create a load of 450,000 pounds so our field factor of safety is 4.7. With regard to modern loads, the 1880 wheel loads are 24,000 pounds or a E24 in more modern nomenclature. Our ratio of train loads are E42= 1.75 (42,000/24,000) which reduces the factor of safety to 4.7/1.75 = 2.7, E72= 3, E80= 3.3: Factor of Safety = 1.4, and E90 = 3.75: factor of Safety = 1.25. Wanting to preserve a safety factor of at least 2 leads to the conclusions as noted by the students.

The project is only approximate but shows how principles from the book can be used for projects to learn engineering principles which relate to real field problems.

This project and the Advanced Projects are pretty hard. If one is completed make sure to award a Certificate of Achievement. (See next page)

Terms used in this book:

A - Area of a section, units $inches^2$
b - base of section
c - distance to the outside fiber from the centroid of a member
cup - English unit
d - distance, units feet or inches
D - Diameter
E- Young's Modulus of a section, units lbs/in^2
F - Force, units pounds or kips
feet - 12 inches
gallon - English measure of volume
h - height of a section
I - Moment of Inertia
inch - distance
K - Stiffness, used as spring stiffness or stiffness matrix
k, kip - 1000 pounds
kilonewton - metric measure of force
l,L - length, units feet
M - Moment (Force times distance), units lb-in
P - Load, units pound or kips
pound, force
psf - pounds per square foot
psi - pounds per square inch
quart - English measure of volume 4 quarts = 1 gallon
w - distributed load of constant value
y - deflection, units inches
Π - pi, constant = 3.1416

The major principles:

1. Spring Law:

   $Ky = P$

   y = deflection, P = load, Constant K relates P and y in a linear manner

   Or in Finite Element Method: Stiffness of a structure, K, subjected to a load, P, can be used to solve for displacements, y, and subsequently solved for loads within members.

2. Normal stress

   σ - stress =P/A

3. Normal strain

   ε - strain, change in length/original length

4. Hooke's Law

   Hooke's Law relates the factors of stress and strain

   Stress = Stiffness * Strain

   Where stiffness is a constant named Young's Modulus

5. Bending stress

   Stress created by a beam in bending, derived from the Deflection Equation as

   Stress = Mc/I

   Where: Terms M, c, and I are defined above

Certificate of Achievement

awarded to:

____

For Completion of an Advanced Study in Trusses

Signed By: ____________________ Date: ____________________

Truss Fun

Answers to selected Problems

Chapter 2 Drawing some trusses:

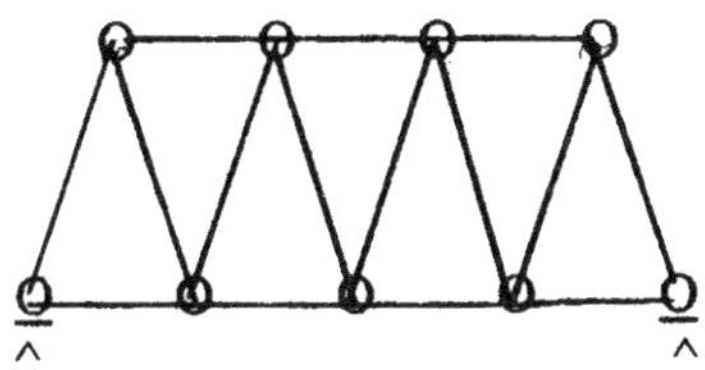

Warren Truss

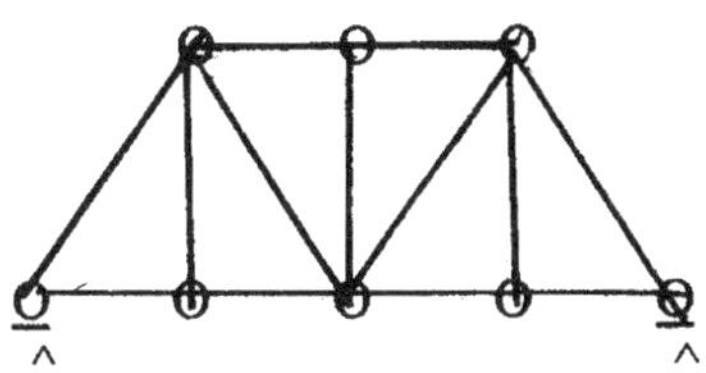

Pratt truss

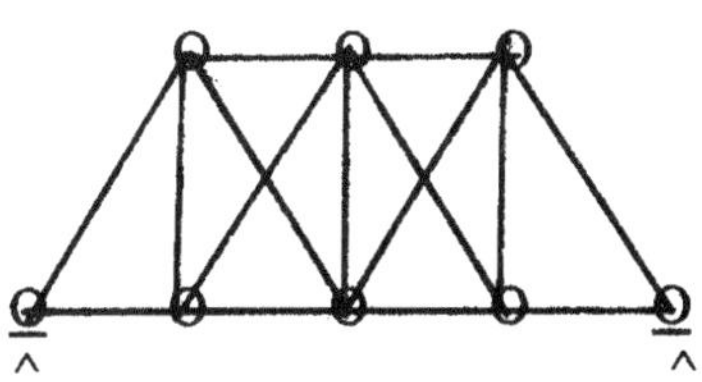

Howe truss

These trusses can only allow trains to enter on the lower chord, these would be through trusses.

Truss Fun

Chapter 5 Geometric Scale:

Problem 1.

Say you measure yourself and you are 5' 1" tall or 61" tall.
The scale would be 61/6 approximately = 10.
Scale is 1 in 10 or 1/10.

Problem 2.

6" is 87 times larger in full scale, 6x87= 523" or 43.5 feet

Problem 3.

6"/ (1/2) =12 Scale is 1 in 12 or 1/12

Problem 4

E40 train wheel loads: 20 kips + 40*4 kips + 26*4 kips = 284 kips

"G"	1/24	11,833 lbs
"O"	1/48	5,916lbs
"HO"	1/87	3,264 lbs
"N"	1/160	1,775 lbs

Models would be too heavy if built to these weights.

Problem 5

Cross brace is "C" section 8" by 20 lbs/ft
Use Table 5.2 Closest is C 8" x 18.75 Model stock 5/16" by 3/32"

Chapter 6 - Analyzing Trusses

Truss Problem

By moments or symmetry Reactions are 40 kips

Pin A Sum F in y direction = 0; $4/5F_{AD}$ + 40 kips = 0; F_{AD} = -50 kips =50 kips Compression;
Sum F in x = 0 3/5(-50 kips)+ F_{AB} =0 F_{AB} = 30 Tension
Pin B Sum F in y = 0 -80 kips + F_{BD} =0 F_{BD} = 80 Tension
All other forces are symmetrical

Spring Problems

Problem 1

K y = P
K(1/8") = ½ pound
K = 4 lbs/in

Problem 2

K y = P
Weight for 1 system = 125,000 lbs/4 = 31250 lbs
Weight for 1 spring = 31250 lbs/4 = 7812. Lbs
K (1 in.) = 7812 lbs.
K = 7812 lbs/in

Simultaneous equations problems

Problem 1a

Solve for x_1 $x_1 = 25 - x_2$ and substitute into equation 2: $2(25 - x_2) - x_2 = 5$
Simplify to find: $-3x_2 = -45$, so $x_2 = 15$, and with equation 1 $x_1 + 15 = 25$, so $x_1 = 10$.

Problem 1b.

|1 1| = | 25| Step:1 Row 1x-2 | 1 1| = | 25 | Step 2: | 1 1| = | 25|
|2 -1| = | 5| add to row 2 |0 -3| = | -45| Row 2x-1/3 | 0 1| = | 15|

Finally: | 1 0| = | 10|
|0 1| = | 15|

Problem 2a.

Solve for x_1: $x_1 = 7/2 - 1/2x_2 - 1/2x_3$ Substitute into equation 2 eliminating x_1 ,or
$(7/2 - x_2/2 - x_3/2) + x_2 + x_3 = 8$, which simplified: $x_2/2 + x_3/2 = 9/2$, or $x_2 + x_3 = 9$, and $x_2 = 9 - x_3$.
Make all substitutions into equation 3 and solve for x_3: $5(7/2 - 1/2\,x_2 - 1/2\,x_3) + x_2 + 2\,x_3 = 8$,
or $-3/2x_2 - 1/2x_3 = -19/2$, and substituting for x_2,
$-3/2(9 - x_3) - 1/2x_3 = -19/2$, $x_3 = 4$, which follows that: $x_2 + 4 = 9$, or $x_2 = 5$, and $2x_1 + 5 + 4 = 7$, or $x_1 = -1$.

With linear algebra,
2b:| 2 1 1| | 7| Row 1 |1 ½ ½| | 7/2| Row 1 x |1 ½ ½| |7/2|
| 1 1 1| = | 8| x ½ |1 1 1| = | 8| -1 add to 2 | 0 ½ ½| = |9/2|
| 5 1 2| | 8| |5 1 2 | | 8| | 5 1 2| | 8 |
Followed by
Row 2 | 1 ½ ½| |7/2| Row 1 | 1 ½ ½| |7/2 | Row 2 | 1 ½ ½| | 7/2|
x 2 | 0 1 1| = | 9| x -5 | 0 1 1| = | 9 | x 3/2 |0 1 1| = | 9 |
| 5 1 2| | 8 | add to3 | 0 -3/2 -1/2| |-19/2 | add to 3 | 0 0 1| | 4 |

Which leads |1 0 0| |-1|
To by | 0 1 0| = | 5|
Substitution | 0 0 1| | 4|

Chapter 7 - Strength of Materials

Problem 1

A bag of sugar is approximately 4" x 5"
Area = A = 4x5 = 20 sq. in.
Lbs/sq in = 5 lbs/ 20 sq in. = 0.4 lbs/sq in

Problem 2

Stress=P/A
Area = A = $3.14(6)^2/4$ = 28.26 sq. in.
Stress = 140,000 lbs/ 28.26 sq-in
Stress = 4.954 lbs/sq-in
Strain = Change in length/Initial length
Strain = 0.02 in./12 in
Strain = 0.0016
Stress = E * Strain
4954 lbs/sq-in = E * 0.0016
E = 3,096,250 lbs/sq-in

Problem 3

Stress=P/A
36,000 lbs/sq-in = P/0.196 sq. in
P= 7,056 lbs.
Stress = E* Strain
36,000 = 36,000,000* strain
Strain = 0.001
Steel will yield, ductile materials
Strain large amount before failure

Problem 4

Stress=P/A
P=8600/4 = 2150 lbs
A= 9x9=81sq-in
Stress= 2150/81 = 26.5 lbs/sq-in
P+ = 6000/4
Stress+ = 18.5 lbs/sq-in
Stress = 26.5+18.5 = 45 lbs/sq-in
65 is greater than 45 or tire would be flat, (Inside pressure is greater than outside stress)

Problem 5

From Chapter 7 $M = -wl/2*(x/2) + wx^2/2$; $x/2=l/2$
So $M=-100(30)/2 - 100(15)^2/4$
M = -22,500lb-ft + 11250 lb-ft = -11,250 lb-ft

$I = 1/12\ b\ h^3$ $I = 1/12\ (12")(24")^3 = 13824\ in^4$

Stress= Mc/I = -11,250lb-ft(12in/1ft) (12")/ 13824
Stress -117.2 psi.

Workbook: Fun for Students - Projects and Problems

A table of historic bridges by state is contained on the following pages. The table is generated from data from states, particularily the state Departments of Transporation, the Library of Congress, and some other InterNet sources. The columns have the headings as defined:

State:

The state of origin of the truss bridge

Bridge:

The name of the bridge if it is given locally

Location:

The county within the state

The city or town where the bridge is located, or may e near (in the vicinity)

A road or location for the bridge

River/Site

The crossing of the bridge, usually a river, creek, or body of water

Truss Type

The construction type of the bridge

Spans

Number and/or length of the spans of the bridge

Const. Date

Construction Date of the bridge

Use:

Intention, use, of the bridge

Every effort was made to find bridges in every state and to tabulate complete information. Some fields were necessarily left blank due to lack of information. Have fun finding and learning about the trusses in your area. History and engineering are fun.

State	Bridge:	Location	River/Site	Truss Type	Spans	Const. Date	Use:
AL		Talladega County Eastaboga Vicinity County Rd. 93	Choccolocco Creek	Lattice (Town)	1	Pre Civil War	
AL	Cripple Creek Covered Bridge	Colbert County Allsboro vicinity County Rd.1	Cripple Creek River	Town	2	1859	
AL	Buzzard Roast Covered Bridge	Colbert County Cherokee vicinity Old Memphis Road		Lattice (Town)	4	1850's	
CA		District 2	Walker Creek		1	1880	
CA	Bay Bridge	SanFrancisco-Oakland				1936	RR and Highway to 1959
CA	Bridgeport Covered Bridge	Nevada County Bridgeport Vicinity	Yube River	Howe	1@223 ft.	1861	
CA	Knight's Ferry Covered Bridge	Stanislaus County Knight's Ferry East of Oakdale	Stanislaus River	Warren	4	1862	
CA	Historic Highway Bridges of California						
CO	Denver S. Park & Pacific Railroad	Chaffee County, Romley					Railroad
CO	Wynkoop RR Bridge	Denver, CO Near Wynkoop St.	Cherry Creek				Railroad
CT		New London County Colchester	Blackledge River	Double Intersection Warren		1907	Railroad
CT	Singing Bridge	Westbrook Route 1	Pachaug River		1@120 ft.		Vehicular
CT	Washington Ave.	New Haven County Waterbury Washington Ave.	Mad River	Lenticular		1880	Vehicular
DE	Smith's Bridge	New Castle County Granogue Smith Bridge Road	Brandywine Creek	Burr	139'	1839	
FL	Sulphur Springs	Pasco County	Hillsborough River				
GA		Cobb County (near) Atlanta Smith Bridge Road	Brandywine Creek	Town	130"		
ID		Fruitland US 95 towards Payette		Steel			Vehicular
IL	Butzow Bridge	Iroquis County Cresent City Vicinity County Hwy 35	Spring Creek	Pratt through		1883	Vehicular
IL	Coal Creek Bridge	Fulton County Fairview Vicinity State Route 97	Coal Creek	Warren Pony		1936	Vehicular
IL	Deer Creek Bridge	Wayne County Geff Vicinity Township Rd. 406	Deer Creek	Warren Pony		1910	Vehicular

State	Bridge:	Location	River/Site	Truss Type	Spans	Const. Date	Use:
IL	Hazen Bridge	Champaign County Mahomet vicinity Township Rd. 85	Sangamon River	Pratt through		1893	Vehicular
IL	Hill Creek Bridge	Pike County Pearl State Route 100	Hill Creek	Warren Pony		1934	
IL		Will County Romeoville Romer Road	Ship Canal	Warren through		1899	Vehicular
IL	Jack's Mill	Henderson County Oquawka vicinity Bridge Road	Henderson Creek	Lattice (Town)	3 for 116'	1846	
IL	Eames	Henderson County Oquawka vicinity State Highway 94A	Henderson Creek	Howe	104'	1861	
IN		Huntington County Huntington vicinity Satamonic Rd.	Wabash River	Pratt through		1885	
IN		Wells County Vera Cruz vinicity State Rt. 316	Wabash River	Double Intersection Pratt			
IN	Mill Creek Bridge	Crawford County Alton vicinity	Mill Creek	Modified Pratt		1885	
IN	Tippecanoe River Bridge	Fulton County Rochester vicinity State Rt. 25	Tippecanoe River	Modified Pratt - Through		1890	
IN	Carroll County Bridge No. 119	Carroll County Camden Vicinity County Rd. 300 W	Deer Creek	Pratt through		1888	
IN	Gospel Street Bridge	Orange County Paoli S. Gospel St.	Lick Creek	Pratt		1880	
IN	Ceylon Covered Bridge	Adams County Geneva County Rd. 900 S	Wabash River	Howe	1@84'	1860	
IN	Adams Mill	Carroll County Cutler County Rd. 50 E	Wildcat Creek	Howe	138'	1871	
IN	Cumberland Covered Bridge	Grant County Mathews County Rd. 990E	Mississinewa River	Howe	167'	1877	
IN	Vermont Covered Bridge	Howard County Kokomo Highland Park	Kokomo Creek	Howe	1@94'	1874	
IN	Medora Covered Bridge	Jackson County Medora State Rt. 235	White River (East fork)	Burr	434'	1875	
IN	Bells Ford	Jackson County Seymour State Rt. 258			2for330'	1869	
IN	Decos Mill	Montgomery County Alamo State Rt. 234	Sugar Creek	Burr	2 for 319'	1878	
IN	Jackson Covered Bridge	Parks County Bloomingdale West of Rt. 47	Sugar Creek	Burr	1 for 207'	1861	
IN		Parks County Mansfield	Big Racoon Creek	Burr	2 for 277'	1867	
IN	Leatherwood Station	Parks County Montezuma	Leatherwood Creek	Burr	1 for 72'	1899	

State	Bridge:	Location	River/Site	Truss Type	Spans	Const. Date	Use:
IN	Narrows	Parks County Turkey Run St. Park County Rd. 280E	Sugar Creek	Burr	1 for 120'	1882	
IN	Busching	Ripley County Versalles Covered Bridge Rd.	Laughery Creek	Howe	1 for 170'	1885	
IN	Kennedy	Rush County Rushville State Rd. 44	Flat Rock River	Burr	1 for 165'	1880	
IN	Wagon Bridge	Union County Brownsville Eagle Creek Park	Whitewater River	Long		1840	
IN		Union County Dunlapsville Roseburg Road	Whitewater River (E. Fork)	Burr		1870	
IA	Eureka	Winneshiek County Frankville	Yellow River	Bowstring			
IA		Warren County Carlisle 3 mi. NE of Carlisle	North River	Howe	100'		
KS		Dickinson County Enterprise K-43 Highway	Smokey Hill River	Parker		1925	Vehicular
KS		Levenworth County Springdale	Stranger Creek	Howe		1859	
KY		Harrison County Berry KY State Rt. 1032	Licking River (S. Fork)	Pratt		1906	Vehicular
KY		Logan County Lickskillet KY Rt. 5273	Whipperwill Creek	Pratt		1880	Vehicular
KY		Harrison County Cynthiana Leesburg Pike	S. Fork Licking River	Burr	2for259'	1808	
KY		Pendleton County Butler	Licking River	Double Burr Truss	3for460'	1870	
ME		Cumberland County South Windham Harry Cane Rd.	Presumpscrt River	Queen	72'	1860	
MD		Howard County Savage	Little Patuxent River	Bollman		1850	Railroad
MD	Waverly St. Bridge	Allegany County Westernport Waverly St.	George's Creek	Bowstring Arch		1892	
MD		Hartford County Jerusalem Jericho Road	Little Gunpowder Falls	Burr	90'	1858	
MA		Hampshire County Northhampton		Lattice		1887	Railroad
MA	Upper Pacific Mills Bridge	Essex County North Andover				1864	Vehicular
MA		Hampton County Ludlow-Wilbraham ENE of Springfield US 20	Chicopee River			1852	

State	Bridge:	Location	River/Site	Truss Type	Spans	Const. Date	Use:
MA	Nehemiah Jewett's Bridge	Middlesex County Pepperell	Nashua River		3 for164'	1818	
MA		Worcester County Millville	Blackstone River			1800's	Vehicular
MI	Trunk Line Bridge	St. Clair County Kimball Griswald Rd.	Pine River	Pratt		1915	Vehicular
MN	Bresseau Road Bridge	St. Louis County Burnett Vicinity County Rd. 694	Cloquet River	Parker Overhead		1911	
MN		Blue Earth County Skyline Township Rd.	LeSueur River				
MN		Ramsey County St. Paul Selby Ave.	Short Line RR Tracks	Pratt		1890	Cable Car Pedistrian
MS		Lowndes County Steens	Yellow Creek	Town		1885	
MO		Benton County Fairfield State Hwy. 7	Usage River				Vehicular
MO		Clinton County Trimble	Lin Branch Creek	Waddell A		1898	Railroad
MO		Cape Girardeau County Allenville	Whitewater River	Town		1865	
MO		Cape Girardeau County Bufordville	Whitewater Creek	Bollinger		1860	
MO	Noah's Ark Covered Bridge	Platte County S. of Edgeton County Rte. B	Little Platte River	Town	120'	1878	
NE		Omaha	Missouri River				Railroad
NE	Platte River Bridge	Butler County Bellwood Township Road	S. Channel of Platte River	Warren Through	73'	1891	Vehicular
NE	Bell Bridge	Cherry County Valentine County Rd 12	Niabrara River	Pratt Through	150'	1902	Vehicular
NE	Bouman Bridge	Cherry County Valentine County Rd (SE)	Niabrara River	Pratt Through	131'		Vehicular
NE	Berry State Aid Bridge	Cherry County Valentine County Rd. (NE)	Niabrara River	Pratt Through	153'		Vehicular
NE	Bryan Bridge	Cherry County Valentine US Hwy 20/83	Niabrara River	Pratt Deck	290'		Vehicular
NE	Brewer Bridge	Cherry County Valentine County Rd. (E)	Niabrara River	Pratt through	152'		Vehicular
NE	Lisco State Aid Bridge	Garden County Lisco S. of Lisco 0.16 m.	N. Platte River	Pratt Pony	5 for 651'	1928	Vehicular
NE	Lewallen State Aid Bridge	Garden County Lewellen	N. Platte River	Pratt Pony	7 for 710'		Vehicular

State	Bridge:	Location	River/Site	Truss Type	Spans	Const. Date	Use:
NE	Gross State Aid Bridge	Knox County Verdigre 3.5 m. North	Verdigres Creek	Parker Through	182'	1918	Vehicular
NE	Browsville Bridge	Nemaha County Brownsville US Hwy 136	Missouri River	Warren through	2 for 1904'	1939	Vehicular
NE	Columbus Loup River Bridge	Platte County Columbus US Hwy 30	Loup River	Parker Through	7 for 1240'		Vehicular
NH		Grafton County Orford Bridge Street	Connecticut River	Town	438'	1850	
NH		Hopkinton E. of NH Rt. 103	Contoocock River	Double Town-Pratt Lattice	2 for 140'	1850	Railroad
NH	Cornish-Windsor Covered Bridge	Sullivan County Cornish On route to Windsor	Connecticut River	Town	2 for 460'	1866	
NJ		Cumberland County Mauricetown	Maurice River	Pratt through		1888	Vehicular
NJ		Somerset County Nehanic Station State Rt. 567	Ranitan River	Lenticular	2 for 280'	1896	Vehicular
NJ		Essex County Newark Rt. 1 and Rt. 9		Parker			
NJ		Hunterdon County Clinton West Main Street	Ravitan River	Lowthrop (after Pratt)		1870	Vehicular
NJ		Burlington County S. Pemberton		King	4 @ 30'	1850's	
NJ	Green Sargent's Bridge	Hunterdon County Sergeantsville	Cold Springs Creek	Howe	84'	1872	
NY		Cattaraugus County Limestone Vicinity		Pratt Through			Railroad
NY		Cattaraugus County Salamanca Vicinity		Baltimore Skewed Through			Railroad
NY		Lawrence County Raymondville Grant Road	Raquette River	Parobolic High			
NY		Genesee County Ratania Vicinity State Rt. 5		Baltimore Through			Railroad
NY		Albany County Normans Kill Vic.		Whipple	110'	1867	Vehicular
NY		Henkimer County Ilion		Queen Post			
NY	Hyde Hall Covered Bridge	Ostego County East Springfield Gummerglass Park	Shadow Brook	Burr	40'	1830	
NY	Blenheim Covered Bridge	Schoharic County North Blenheim		Long	232'	1854	
NY	Perrine's Bridge	Ulster County Rifton Vicinity Rt. NY213	Near Dashville Falls	Burr	138'	1850	

State	Bridge:	Location	River/Site	Truss Type	Spans	Const. Date	Use:
OH	Abbotts	Seneca County Tiffin Vicinity Abbott Road	Sardusky Road	Parker Through	218'	1897	Vehicular
OH	Cleves Bridge	Hamilton County Cleves Vicinity US Hwy 50	Great Miami River	Parker Through		1914	
OH	Falling Rock Camp	Licking County Hickman vicinity	Rocky Ford Creek	Post			
OH	Forder Bridge	Paulding County Antwerp vicinity County Rt. 73		Pratt through	358'	1889	Vehicular
OH	Fosnaugh Bridge	Fairfield County Stoutsville vicinity Township Rt. 128	Scippo Creek		52'	1891	
OH		Hamilton County Miamitown Harrison Rd(Rt.457)	Great Miami River	Camelback Through	464'	1894	Vehicular
OH		Mahoning County Poland Main St. Rt. 170	Yellow Creek	Pony Truss	100'	1901	Vehicular
OH	Mallaham Bridge	Putnam County Pandora Vicinity County Route M-6	Riley Creek	Bowstring Arch		1876	Vehicular
OH	Oak Knoll Park	Stark County Massillon	Oak Knoll Park	Howe		1859	Vehicular
OH		Hamilton County Ross vicinity Route 463	Great Miami River	Pennsylvania Through	368'	1894	Vehicular
OH		Scioto County Portsmouth State Rte. 73	Scioto River	Pennsylvania Through	336'	1915	Vehicular
OH		Knox County Fredercktown Sockman Road	GrannyCreek	Howe		1873	Vehicular
OH		Van Wert County Van Wert Vicinity	Town Creek	Truss-leg Bedstead	44'	1894	Vehicular
OH		Coshocton County Coshocton	Lock on Ohio & Erie Canal	Howe		1933	
OH		Mahoning County Poland Riverside Drive	Yellow Creek	White Bowstring Arch	127'	1877	Vehicular
OH	Zoarville Station	Tuscarawas County Zoarville Vicinity Route 212	Conotton Creek	Fink Through		1868	Vehicular
OH	Station Road Bridge	Cuyahoga Valley National Recreation Area				1881	
OH		Butler County Collinsville Vicinity near US Hwy. 127	Seven Mile Creek	Town	192'	1870	
OH	New London Pike	Butler County Hamilton vicinity State Rte. 27	Indian Creek	Town	76'	1865	
OH		Clinton County Clarksville Route 3	Todd Creek	Lattice	2		
OH	Bright John No. 2	Fairfield County Carroll vicinity Bish Road	Poplar Creek	Burr	72'	1881	

State	Bridge:	Location	River/Site	Truss Type	Spans	Const. Date	Use:
OH		Muskinggum County Zanesville vicinity Old National Trail	Muskingum River			19th Century	
OH		Trumbull County Newton Falls Broad Street	Mahoning River	King		1808	
OH		Union County North Lewisburg State Rte. 275	Spain's Creek	King	64'	1859	
OH		Union County North Lewisburg State Rte. 275	Big Darby Creek	Howe	94'	1859	
OR		Lane County McKenzie Bridge Horse Creek Road	Horse Creek Road	Howe	105'	1930	
OR	Jordan Covered Bridge	Linn County Scio vicinity	Santiam River	Howe	96'	1937	
PA	Allenwood Bridge	Northumberland Coun Dewart vicinity State Rt. 44	Susquehanna River	Pratt through			
PA		Blair County Altoona RailroadersMuseum	1300 Ninth Ave.	Haupt Truss			
PA	Kuhn's Fording Bridge	Adams County East Berlin vic. Township Rd. 552	Conewago Creek	Pratt	147'	1897	
PA	Synder's Fording Bridge	Adams County Hunterstown vic.	Conewago Creek	Burr	2for 147'	19th Century	
PA	Griesemer Mill	Berks County Spangsville Spangsville Road	Manatawny Creek	Burr	124'	1832	
PA	Bartram's Covered Bridge	Chester County White House vicinity Chester Rd15098	Crum Creek	Burr	2 for 80	1860	
PA		Erie County Waterford	LeBoeuf Creek	Town	65'	1880	
PA	Grimes Covered Bridge	Greene County Ruff Creek Vicinity US 19	Ruff Creek	King	34'	1888	
PA	Stoneroads Mill Bridge	Lancaster County Maple Grove vinicity	Little Conestoga Creek	King	65'	1868	
PA	Leaman Rifle Works Bridge	Lancaster County Pinetown	Conewago Creek	Burr	67'	19th Century	
PA	Johnson's Mill Bridge	Lancaster County Rapho-Hempfield	Little Chickie's Creek	Burr	80'	1866	
	Thomas Mill Covered Bridge	Philadelphia County Philadelphia Thomas Mill Road	Wissahickon Creek	Burr	96'	19th Century	
PA	Gross Covered Bridge	Synder County Beaver Springs Township Road 574	Middle Creek	Burr	100'	1878	
PA	Bell's Mill Bridge	Westmoreland Cnty West Newton vic. US 71	Sewickley Creek	Burr	104'	1850	
PA	Detter's Mill Covered Bridge	York County Detter's Mill Harmony Grove Rd	Big Conewago Creek	Burr	258'	1815	

State	Bridge:	Location	River/Site	Truss Type	Spans	Const. Date	Use:
TN	Elizabethton Covered Bridge	Carter County Elizabethon Hattie Ave.	Doe River		134'	1882	
TX	Fort Griffin	Shackelford County Albany vicinity CR 188	Clear Fork of Brazos River			1885	
UT		Washington County Hurricane vicinity St. Hwy. 9	Virgin River	Warren			
VT		Caledonia County St. Johnsbury Mount Vernon St.	Sleeper's River	Lenticular		1885	
VA	Survey of Metal Truss Bridges in Virginia						
VA		Rockingham County Broadway State Rt. 1414	Dahna Creek	Pratt	68'	1900	
VA		Augusta County Crimora vicinity State Rt. 612	South River	Pratt Through			
VA		Rockingham County Broadway	Linville Creek	Thacher		1898	
VA		Augusta County Cave vicinity State Rt. 256	Middle River	Pratt Through			
VA		Bedford County Bedford vicinity State Rt. 657	Roaring Run	Bowstring		1877	
VA		Allegheny County Greenwood vicinity VA Route 60	Dunlops Creek	King	109'	1835	
VA	Marysville Covered Bridge	Campbell County Gladys vicinity State Road 633	Seneca Creek	Multiple King	60'	1878	
VA	Trent's Bridge	Cumberland County Cumberland vicinity	Willis River	Town	145'	1844	
VA		Rockbridge County Lexington In town	North River	Burr	4	19th Century	
WA		Wahiakum County Gray's River vic. Worrel Road	Gray's River	Howe	148'	1905	
WV		Cabell County Milton	Mud River	Howe	112'	1876	
WV	Statt's Mill Covered Bridge	Jackson County Ripley vicinity Secondary 40	Tug Fork River	Long	97'	1887	
WV		Marion County Barracksville Secondary 145	Buffalo Creek	Burr	132'	1853	
WI		Ozaukee County Cedarburg vicinity Five Corners Road	Cedar Creek	Town	120'	1876	
WY		Sheridan County Arvada vicinity County Rd. 38	Power River	Parker Through	160'	1917	Vehicular
WY	Bessemer Bend Bridge	Natrona County Casper vicinity County Rd. 8	N. Platte River	Warren Through	185'	1922	Vehicular

State	Bridge:	Location	River/Site	Truss Type	Spans	Const. Date	Use:
WY		Sheridan County Sheridan vicinity County Rd. 51	Big Goose Creek	Pratt Pony		1914	Vehicular
WY	Big Island Bridge	Sweetwater County Green River vic. County Rd. 4	Green River	Pratt Through	310'	1910	Vehicular
WY		Fremont County Dubois vicinity County Rd. 21	Big Wind River	King Post	73'	1920	Vehicular
WY		Union County Fort Badger vicinity County Rd. 217	Black's Fork River	Warren Pony	80'	1925	Vehicular
WY	Butler Bridge	Carbon County Encampment vic. County Rd. 203	N. Platte River	Camelback Through	170'	1930	Vehicular
WY		Niobrara County Riverview vicinity County Rd. 46	Cheyenne River	Pennsylvania Through	130'	1915	Vehicular
WY	Four Mile Bridge	Hot Springs County Thermopolis vic. County Rd. 173	Big Horn River	Pennsylvania Through	175'	1928	Vehicular
WY		Sweetwater County Green River vic. County Rd. 8	Green River	Pinned Pratt Through	150'	1913	Vehicular
WY	Irigary Bridge	Johnson County Sussex Vic. County Rd. 254	Powder River	Pennsylvania Through	250'	1913	Vehicular
WY		Platte County Wheatland vic. County Rd. 204	Laramie River	Pratt Pony	71'	1913	Vehicular
WY		Albany County Bosler Vic. County Rd. 740	Laramie River	Pratt Half Hip	2@75'	1926	Vehicular
WY		Carbon County Elk Mountain vic. County Rd. 120-I	Medicine Bow River	Warren hybrid		1924	Vehicular
WY		Sublette County Boulder vicinity County Road 136	New Fork River	Kingpost	2@46'	1917	Vehicular
WY		Goshen County Fort Laramie vic.	N. Platte River	Bowstring	420'	1875	Vehicular
WY		Hot Springs County Thermopolis vic. County Rd. 287	Owl Creek Bridge	Pinned Camelback	124'	1920	Vehicular
WY		Sheridan County Leiter vicinity County Rd. 269	Powder River	Pratt and Warren		1915	Vehicular
WY		Sheridan County Arvada vicinity US Hwy 14/16	Powder River	Pratt Deck	3 for 452'	1933	Vehicular
WY		Big Horn County Shell vicinity County Rd. 57	Shell Creek	Warren Pony	60'	1920	Vehicular
WY		Big Horn County Lovell vicinity County Rd. 111	Shoshone River	Warren Pony	4for 380'	1926	Vehicular
WY		Johnson County Kaycee Vic. West if I-25	Powder River	Pratt	3 for 306'	1932	Vehicular
WY		Sheridan County Monarch vic. County Rd. No.97	Tongue River	Pratt	80'	1913	Vehicular

The Newspaper Bridges

Do you think it is possible to build a small bridge from rolled newspaper that you can actually walk across? It is a challenge, but amazingly, it can be done. This section shows other types of bridges besides trusses, an InterNet computer program: The West Point Bridge Designer, and example student projects with Newspaper Bridges you can walk across.

ALWAYS BE SAFE. We are building bridges that will probably fail with a big enough load. Be careful of falling hazards, and other safety issues. This project should be treated like gymnastics, or other events, where falling is a possibility. Be careful of glue and other materials, follow all manufacturer's warnings and instructions. LIABILITY RELEASE: BaHa Enterprises, LLC is not responsible in any way for results from attempting these projects.

Types of Bridges

For our purposes, bridges will be described based on two factors: The forces internal in the bridge (and how they are distributed), and the forces placed on the bridge foundation. The internal forces affect our design of the bridge elements or members, while the loads on the foundation affect the design of the placement of the bridge - where it can be placed for us to walk across. Using these distinctions, most bridges can be divided into the following categories:

Beam Bridges
Truss Bridges
Arch Bridges
Suspension Bridges
Cable Stayed Bridges

Beam Bridges

Beam bridges are generally laid or set down across the distance to be spanned. Since the members are beams, each member is usually subjected to compression in the top of the beam, tension in the bottom of the beam, and moment forces throughout the length of the beam. A typical beam bridge with the types of internal and foundation forces overprinted is shown in Figure NB-1. As the bridge is essentially set down on its foundations, forces into the foundation are usually downward only.

Truss Bridges

Truss bridges are, of course, the main topic of this book. For review, the forces in a truss are only along the member or element of the truss. The forces in the members may be either tension or compression, dependent of the loads placed on the truss and its design. As trusses have pinned connections, no moments exist anywhere within the truss. Forces in truss bridges are shown in Figure NB-2. Like the Beam Bridge, the truss bridge is generally set on the foundation, producing only downward forces into the foundation.

Arch Bridges

Arch bridges have internal loads along the arch. Arch bridges were developed in very early history; many bridges built by the romans were arch bridges. Arch bridges can be made of stone placed to form the arch, thus, the forces must be in compression only to assure that pieces can not fall out. The foundation loads are also important to hold the arch together. The forces along the arch internally, have force components both down and laterally in the foundation, see Figure NB-3.

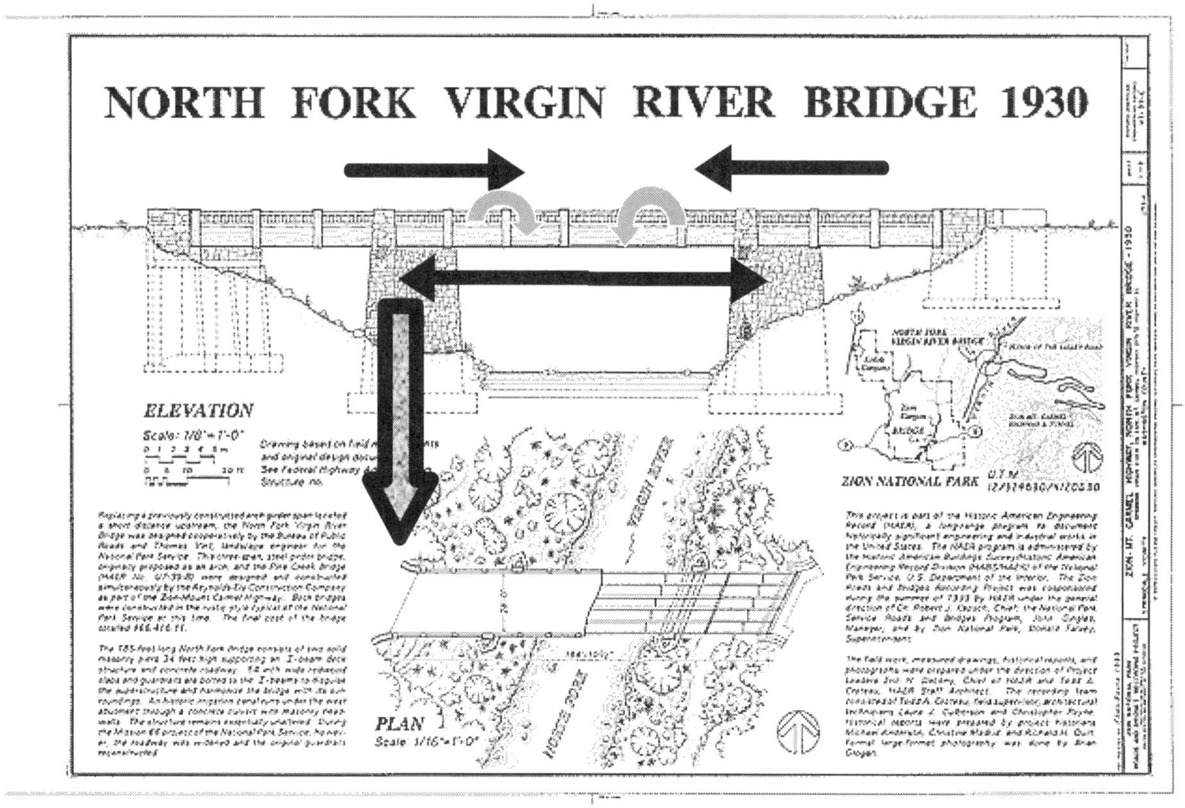

Figure NB-1 Beam Bridge - Internal Forces: Compression- Top, Tension - Bottom, Bending Throughout Length Foundation: Down only.

Figure NB-2 Truss Bridge Internal Forces - Along members Tension or Compression, Foundation Down Only

Figure NB-3 Arch Bridge Internal Forces - Compression Only, Foundation Down and Lateral

Suspension Bridges

Suspension bridges are very interesting and exciting examples of bridges. It was through the use of suspension bridges that longer and longer spans were possible. Such an example is the Wheeling Suspension bridge shown in Figure NB-4. Built in 1869, this bridge spans an excess of 1000 feet in its crossing of the Ohio River. From this figure, it becomes clear that a major cable is stung between two towers and anchored securing into the foundation. Hanging cables are hung from the major cable for the deck of the bridge.

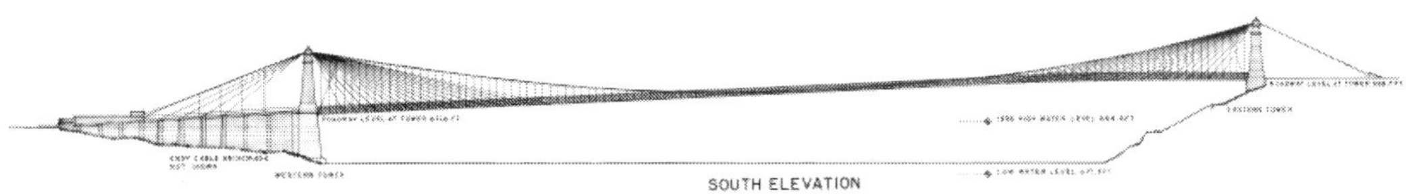

Figure NB-4 Wheeling suspension bridge

More detail for a different suspension bridge is shown in Figure NB-5. Again a major cable is used with hangers for a deck. In these figures the internal loading in the cable is seen to be tension only, compression would only kink the cable. In the foundations for the towers, lateral loads are usually balanced to allow only downward loads. The anchor blocks for the cables support large lateral loads.

Cable Stayed Bridges

Cable stayed bridges are much like suspension bridges but usually have the cable originating from the tower and down to the deck. Thus, supporting cables form large and larger traingles as they radiate from the tower. This design is very stable in wind and earthquake loads owing partially to these triangles in the design.

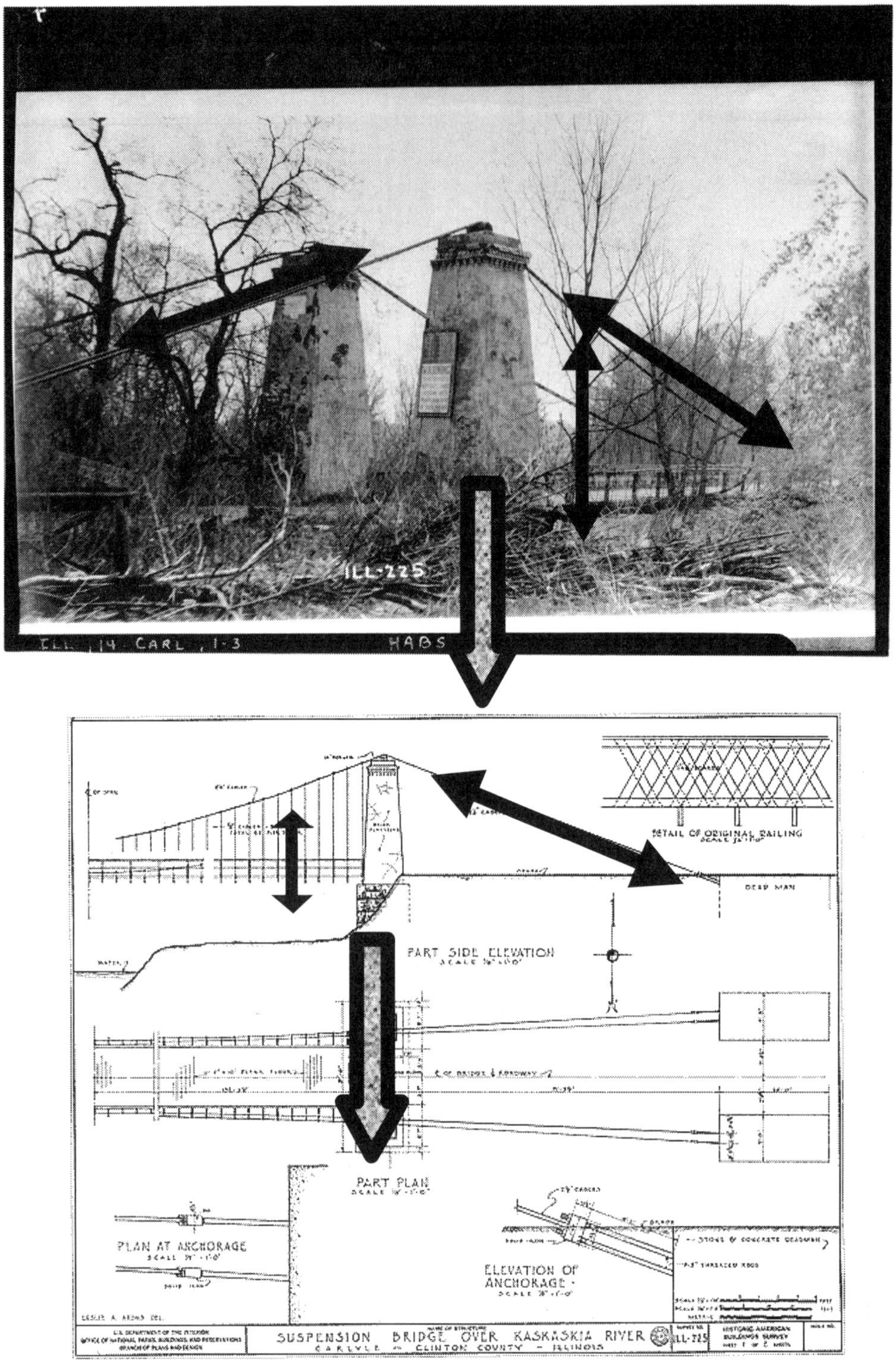

Fighre NB-5 A Pedistrian Suspension Bridge

The Newspaper Suspension Project

For a fun project, a frame may be constructed and used to build any type of bridge. Starting with a frame gives some stability for safety reasons, and is an excellent engineering project. This is because many different types of bridges are candidates for the construction, and the type of bridge can selected which best suits newspaper as a construction material.

Some data showing the load capacity of rolled newspaper from a project done by students (See The Newspaper Truss and Other Newspaper Bridges CD, also by BaHa Enterprises, LLC) is shown in Figure NB-6.

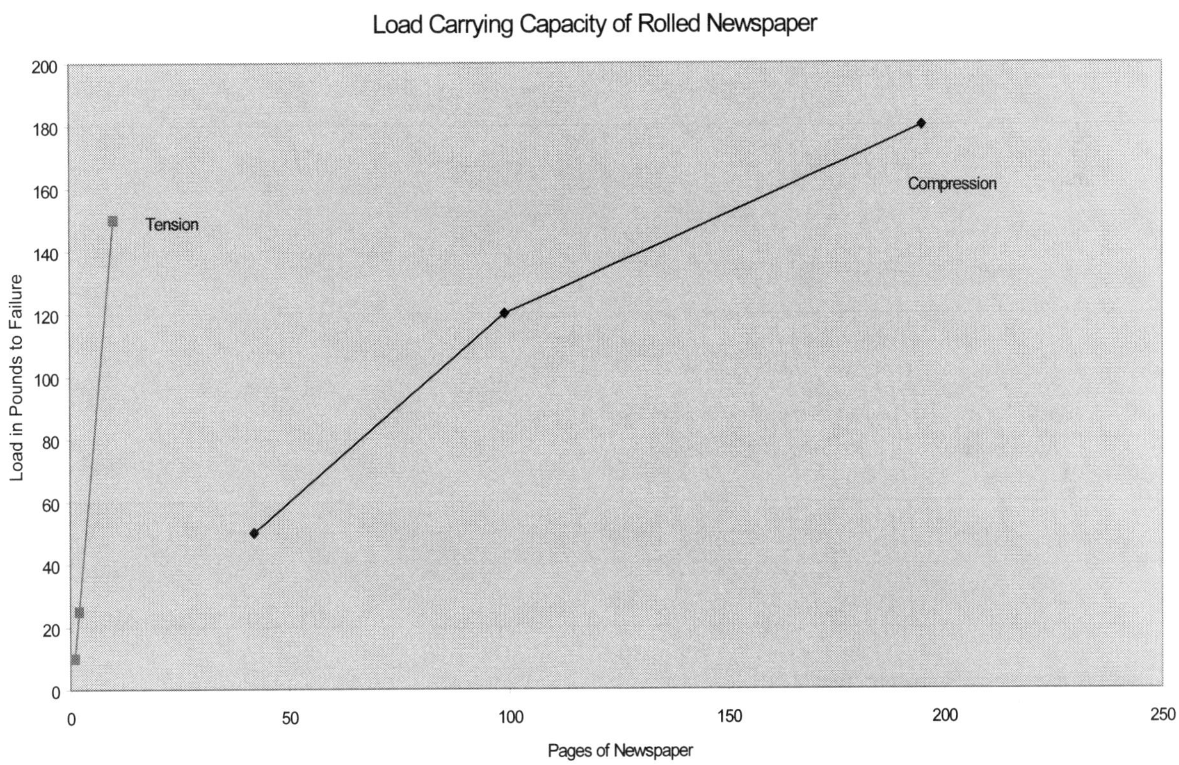

Figure NB-6 The Load Capability of Rolled Newspaper in Tension and Compression

From these data note that the x-axis is the number of pages of rolled newspaper needed to hold a given load. From these few points found by experimentation by the students it can be seen that newspaper will hold much more load in tension than in compression from rolls made of the same number newspaper pages. Thus, if given the choice, members which hold only tension will require smaller rolls and be more efficient, less paper will hold more weight.

A suspension bridge holds only tension in the main cable members. So any type of suspension design will be efficient if towers and anchors are provided as a starting point.

The 2002 Colorado MESA Bridge Design Challenge

In 2002, the Colorado MESA (Mathematics, Engineering, Science Achievement) organization, in cooperation with BaHa Enterprises, LLC suggested a challenge competition for their high school groups to build bridges from newspaper that the students could walk across. For this competition, believed to be the first of its kind, a frame was suggested as a starting point for the students. The frame was made from ordinary lumber with the dimensions as shown below from the contest rules.

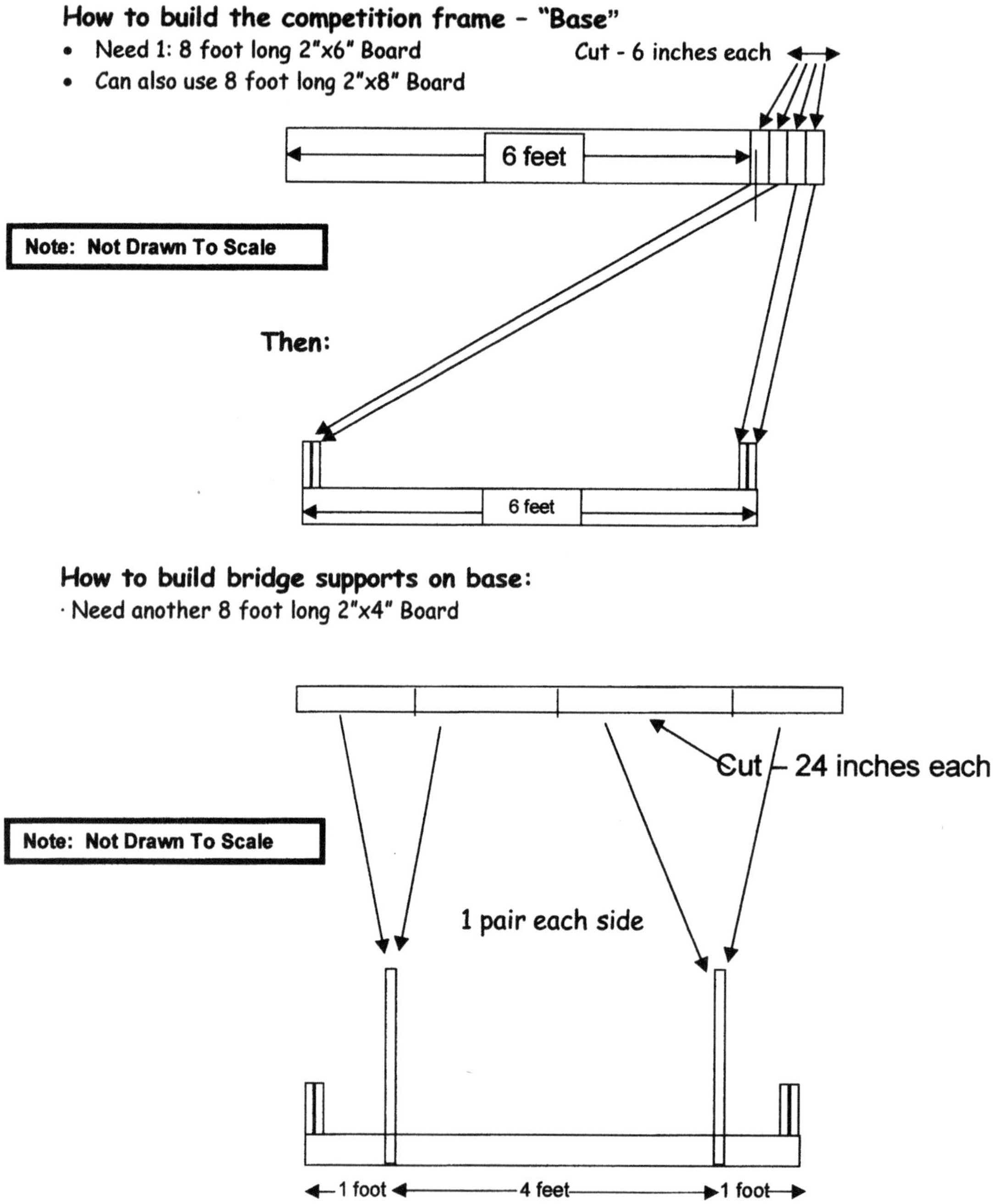

Figure NB-7 Construction details of the 2003 Newspaper Bridge Contest

Final Base

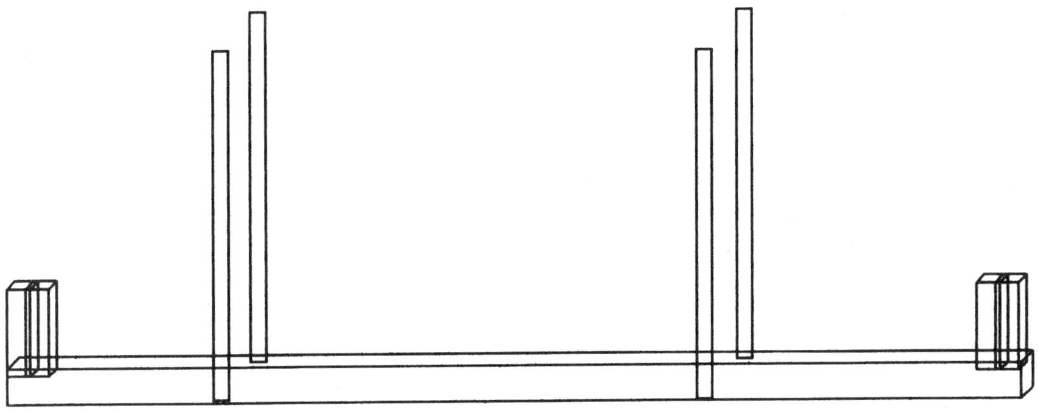

(View 1) Note: Uprights to be placed to walk between

Contest Rules

Objectives of the event are:
1. Build competition frame
2. Design a bridge with materials allowed
3. Build a bridge from newspaper attached to and using the base
4. Walk across
5. Learn some principles of engineering
6. Be safe
7. Have fun

The Rules:
1. Materials:
    a. Newspaper
    b. Not more than 100 screws, not more than 3" in length
       Note: different sizes and kinds can be mixed
    c. Not more than 1 roll of tape (roll typically found in a hardware store)
    d. Not more than 1 container of glue (1 Caulking tube, 1 bottle/can - not to exceed 1 pound)
    e. Not more than 1 gallon of plaster

2. Design:
    Any bridge type is allowed, see Types of Bridges.

3. Evaluation:
    Costs to be provided as part of contest submission (see Table which follows)
    Each person to cross is weighed
    Winner is highest weight of person that crosses in Pounds/Pennies spent
    If no successful crossings, winner based on judges award of innovation, creativity, and best use of design.
    Each crossing is to use spotters to protect the person on the bridge if it does fail. A spotter should be on each side. Be safe!! Think of this as gymnastics or other sports where a participant could fall.

Contest Rules Continued

Disqualification:

1. The use of materials other than newspaper to construct the bridge will be cause for disqualification. Materials from the materials list may be used in combination with newspaper, but may not be the pimary material, e.g. bridge may not be made from 1 roll of strapping tape. Blantant attempts to avoid use of newspaper in construction will be cause for elimination, i.e., **the bridge must be made from newspaper.**

2. Unsafe crossings with no spotters.

Rule additions for the 2004 Contest follow on page 152.

Table for Computation of Costs to Consruct Bridge

CO MESA 2003 Bridge Contest Cost Table
(Bring to Competition Filled Out)

Item	Unit Cost	Number Used	SubTotal Cost
Newspaper (Per sheet)	$0.01		
Screws (per screw)	$0.10		
Tape (any use of roll)	$10.00		
Glue (any use)	$5.00		
Plaster (any use)	$2.00		
Total Cost			

Note: Cost of Bridge Materials Only,
Do NOT include cost of competition base

Newspaper Bridges - More Fun for Students

The teams from Bollman Technical Education Center and Ranum completed their newspaper bridges and walked across during the Colorado High School MESA competition. Some scenes from the competition are shown in Figures NB-9 and NB- 10 respectively.

From Left to Right:
Chris Vallee
Bryce Powell
Scott Reed
Bethany Adams

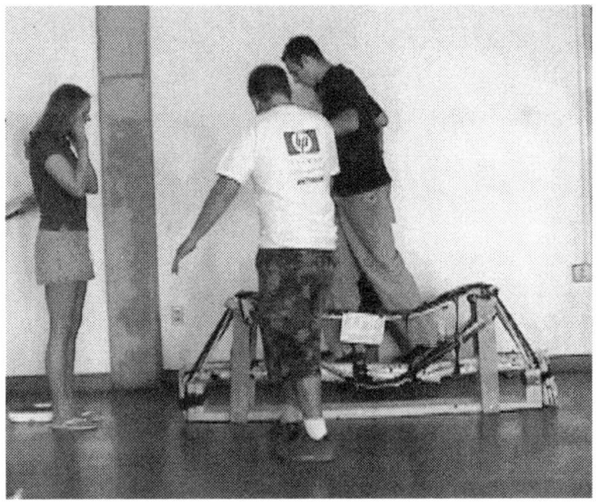

Figure NB 9 The Bollman Tech Newspaper Bridge Team and a Student Crossing - Great Job!

Figure NB-10 Ranum High School Suspension Bridge - Congratulations on a great design of a traditional suspension bridge

The student bridges showed great creativity and very different designs. It was great fun to watch student after student walk across a bridge designed and built from newspaper. Congratulations to all for their engineering accomplishments.

Following the competition some examples of the principles and tools suggested in this book were used to show how to analyze a bridge design. The use of gemeotry and freebody diagrams to make quick and easy hand calculations is shown on page 157 for the Ranum bridge. The use of the Finite Element Computer Program from Chapter 6 is shown on page 158 using the Bollman Tech. bridge.

The 2004 Challenge

For 2004 bridges will be longer and better. We hope you will be interested in this challenging and fun project. Always be safe!! Engineering is about safety too.

Contest Theme:

Our objective is to build a bridge, typical of a pedistrian bridge in a park. Kids, adults, dogs, bikes, etc would be crossing our bridge.

Design additions:

Bridge must have a deck for passage over the bridge. Deck not to exceed 16" from the floor.

New Rules to improve crossings

Each person is to step a minimum of 2 steps in each end span, and 4 steps in the center span. Each crossing will hold 5 seconds at the center of each span. Each crossing is to use spotters to protect the person on the bridge if it does fail. A spotter should be on each side. Note spotters are nearby (a few inches), person crossing does not use the spotter for balance. A crossing which touches a spotter will be disqualified, but can make another attempt.

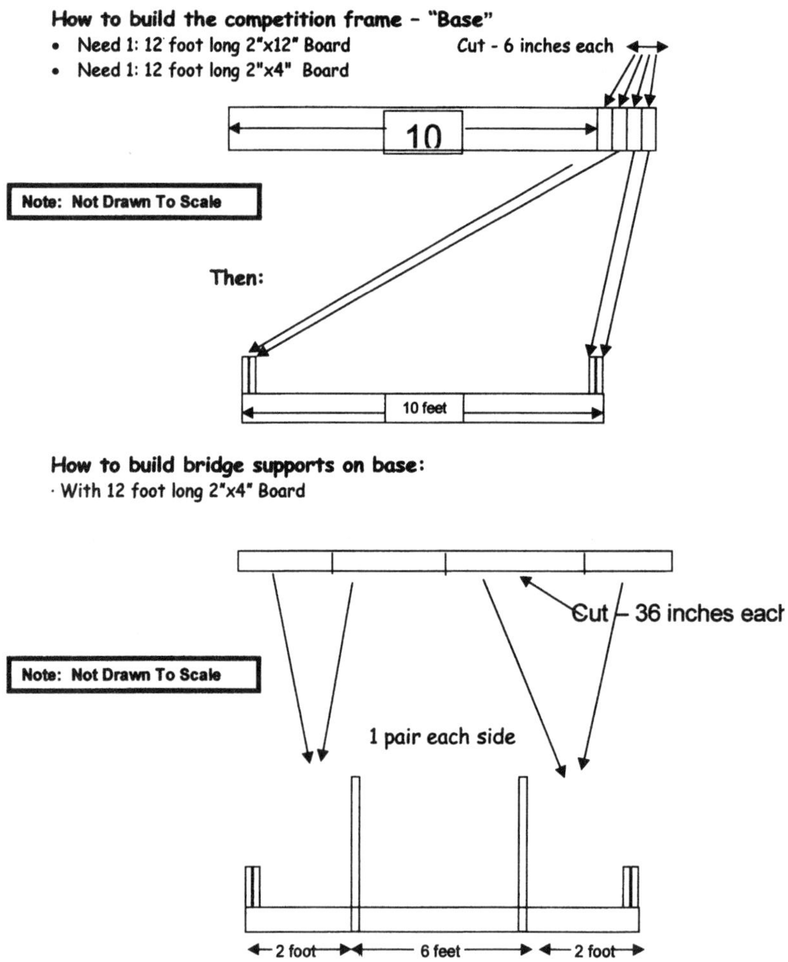

Figure NB-11 Construction for the 2004 Newspaper Bridge Contest

The Newspaper Truss Project

A crossable newspaper truss bridge can also be designed. This project introduces two additional design challenges: 1) the newspaper members are in both compression and tension, and 2) the stability of the bridge must be incorporated into the design. Any truss configuration or design for the truss can be chosen. Figure NB-12 shows one possible design which has been successfully completed by a high school design team and serves as a good starting point.

Figure NB-12 The Crossing of a Newspaper Truss Bridge

To help understand the design needs for the members, the maximum compression and tension forces in the members for a unit load (Load, P=1.0) anywhere in the truss are shown in figure NB-13. Note that the forces are symmetrical about the centerline. Using this design aid tension and compression members are identified, and the member capacity can be estimated, e.g. if our heaviest design team member weighs 150 pounds, our coefficients for the members need to be multiplied by 75 for the load capacity (Note that half of the load is carried on each of the two trusses on the sides). Compression members will require larger members (more sheets), see figure NB-6, or can be designed and constructed as composite members using plaster fill. This allows the design to best use materials; newspaper, strong in tension- weak in compression, and plaster, strong in compression – weak in tension. Thus, the design of the members becomes an integral part of the design.

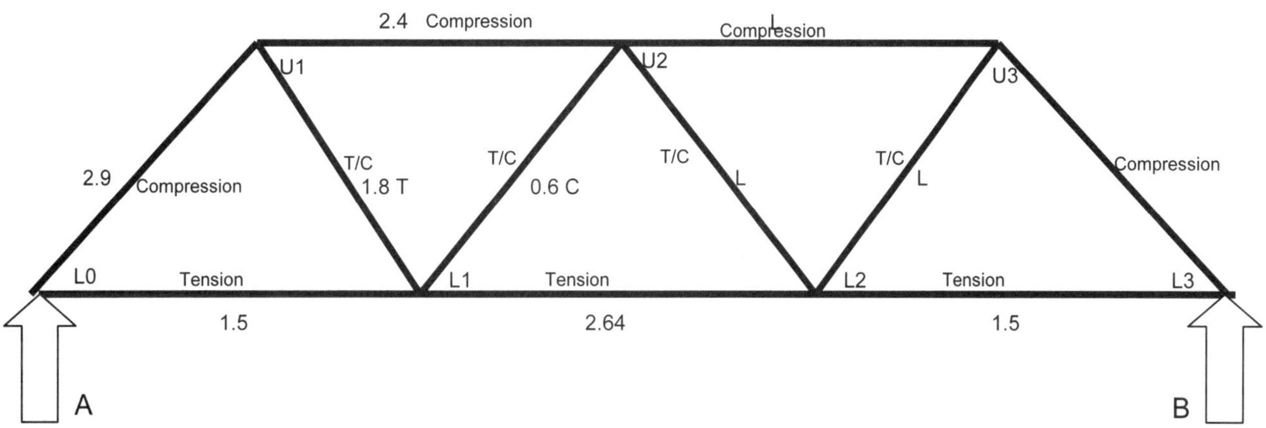

Figure NB -13 Maximum forces in a truss with a unit load placed independently at each truss joint.

Truss Fun

The Colorado School of Mines, High School Newspaper Truss

In 1999, a team of high school students attending the SUMMET (Summer Minority Engineering Training) program at the Colorado School of Mines formed a design team to build and cross the first bridge of this type. The entire project is shown on the CD, "The Newspaper Truss, and Other Newspaper Bridges", also from BaHa Enterprises, LLC, with fun video clips and an entire description by the students of their work. The pasteboard below shows highlights of the project.

Truss Members

A Truss Joint

The First Truss

The Bridge Deck

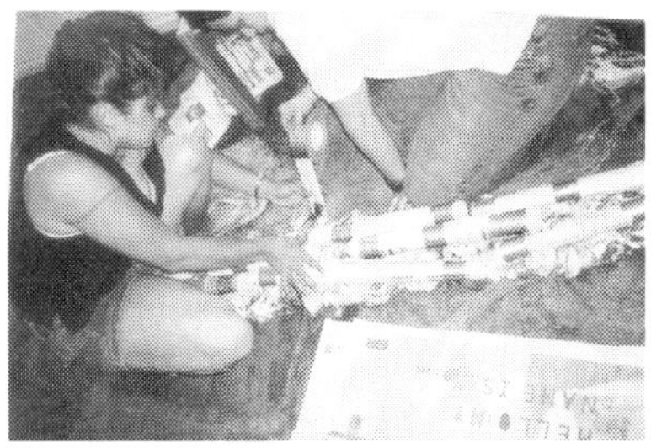

Deck With Strengthening Underneath

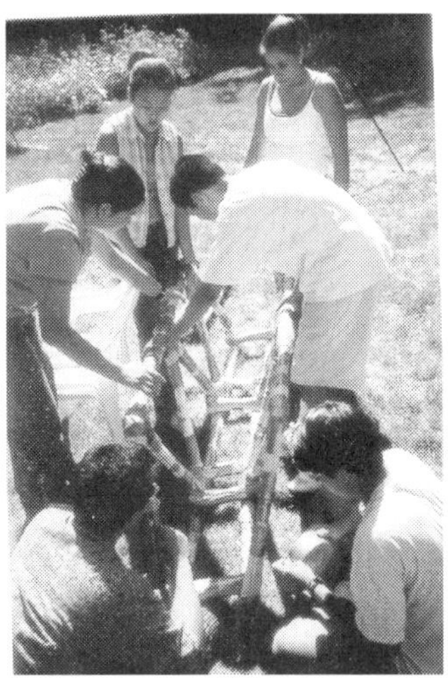

Truss Structural Assembly

Assembly of Pavement

Figure NB-14 The Colorado School of Mines Truss Bridge Project

Newspaper Bridges - More Fun for Students

Colorado Springs 2003 Newspaper Truss Contest

The challenge and successes to build and cross newspaper truss bridges continues. In the Fall of 2003 the Colorado Springs, CO MESA contest included Newspaper Trusses as their high school student's contest. Successful crossings are shown in the photos below.

Figure NB-15a Congratulations to Coronado High School - A successful crossing and the Winner

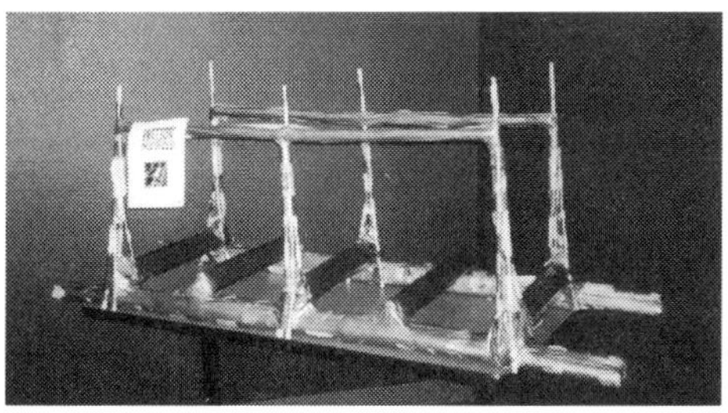

Figure 15b Wasson High School
Another design idea

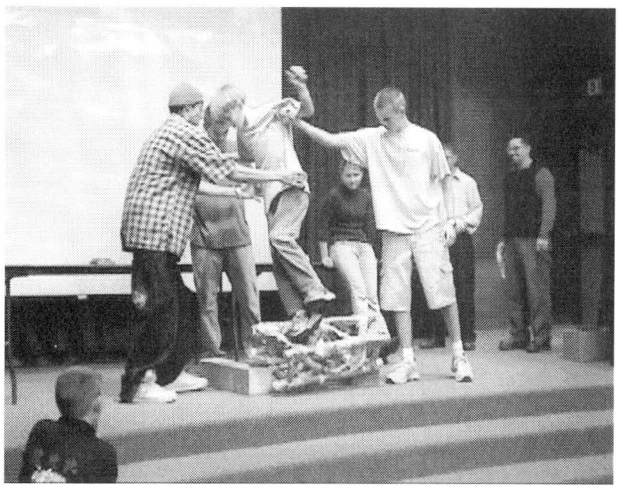

Figure 15c Saved by the Spotters - ALWAYS BE SAFE!!!
Thanks to Doherty High School for this Good Example

Figure NB-15 Colorado Springs 2003 Newspaper Truss Competition

Congratulations to each school for their independent and creative designs. Note the differences in the ideas and concepts. Remember - engineering is a combination of creativity, common sense, and principles learned in science and math.

The West Point Bridge Designer and Contest

The United States Military Academy, West Point, has produced and offers several versions of free, public domain software which can be used to experiment with designs of various bridges, (WPBD4), or can be used to enter an on-line competition, WPBD 2003 (and future editions). For virtual bridge designers around the globe, this software is the video game of bridge design. It can currently be found on the website http://bridgecontest.usma.edu or is readily found with a search engine using "West Point Bridge Designer" as the search text.

West Point has a long history of Civil Engineering education starting in 1802 as the first school in the United States to offer a formal program of instruction in engineering. Academy graduates designed and built many early railroads. The West Point contest commenorates West Point's role in developing the nation's infrastructure. This software provides another aid to understand the principles in this book of applying math, science, and common sense to solving problems with simulated hands-on problems.

The West Point WPBD4 program allows a variety of bridge types to be designed as crossings to fixed river conditions. Bridges for these fixed site conditions may be a number of different trusses, single and multiple span, or may suspension bridges with trusses for deck support. Although it is not possible to use designs of any dimension, this is a fun program and provides an excellent learning opportunity. Bridges are loaded by a truck "driving" across the design and if the bridge is overloaded it actually fails and collapses on screen, a delightful feature.

An example of the use of West Point for our designs is shown if Figure NB- 17 showing the Bollman Tech. bridge fitted to the West Point spans.

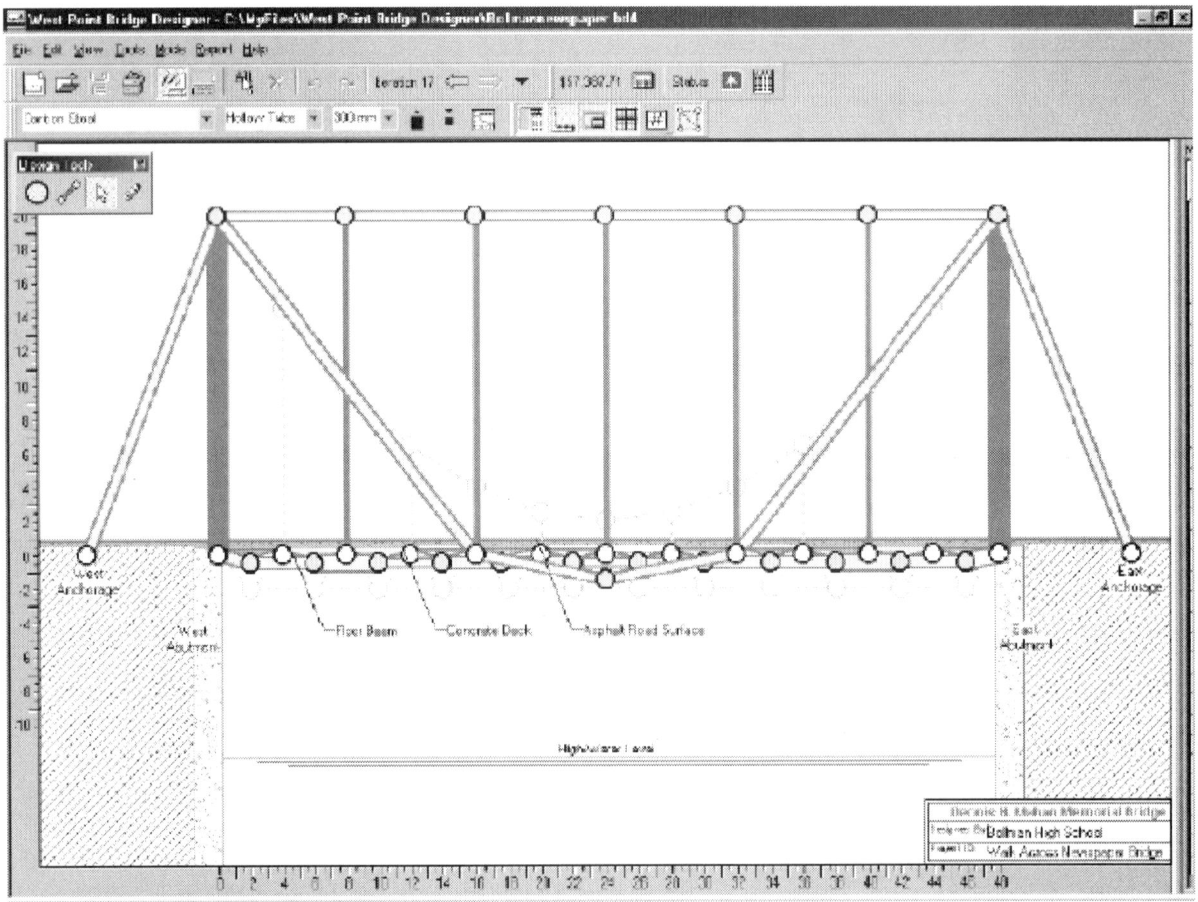

Figure NB-17 The Bollman Tech. Bridge as a West Point Bridge

Use of Geometry to Resolve Internal Forces

Chapter 6 introduced hand analysis methods for forces in bridges. The simple elegance of the Ranum High School bridge design provides an ideal example for hand analysis. Recall that forces in trusses are represented with the same geometric relations as the geometry of the bridge. The approximate geometry of the Ranum bridge is hand drawn, similar to Chapter 6, and shown in figure NB-18.

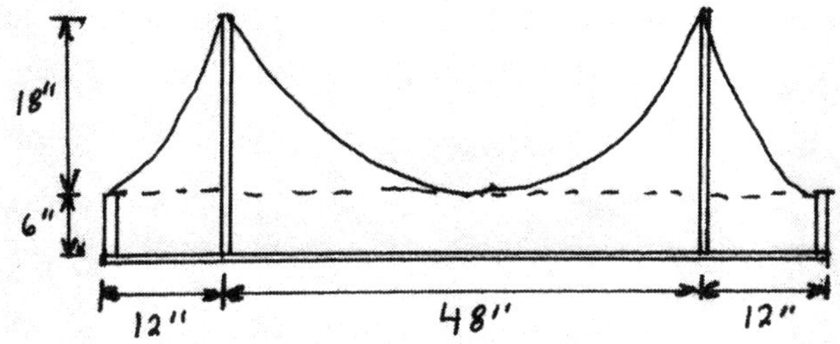

Figure NB-18 Approximate Geometry of the Ranum High School Newspaper Bridge

Figure NB-19 shows a 100 pound force applied at the center of this span and the associated forces. Note that there are 2 sides to the bridge, therefore the 100 pounds shown would actually represent 200 pounds at the center of the span. Although this would be a large load on the bridge, the 100 pound force was chosen as a convenience, since other loads could easily be proportioned from these loads. For example, a 150 pound person would place 75 pounds of each side and so all results found would be multiplied by 75/100 or 3/4.

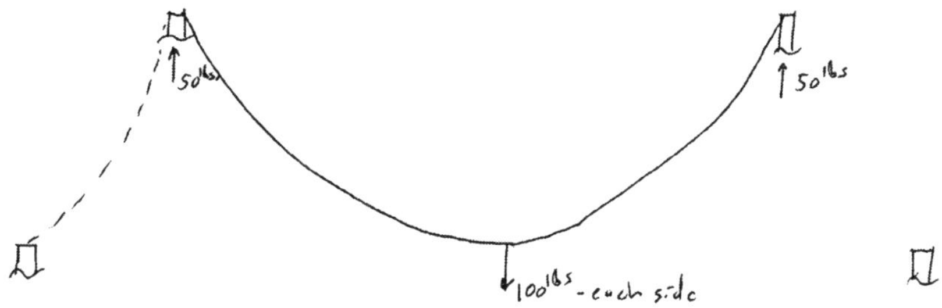

Figure NB-19 Forces on the Ranum High School Bridge

A freebody diagram of internal forces shows that the 50 pound reaction applied to this 3-4-5 triangle produces a force in the suspension cable of 83 pounds.

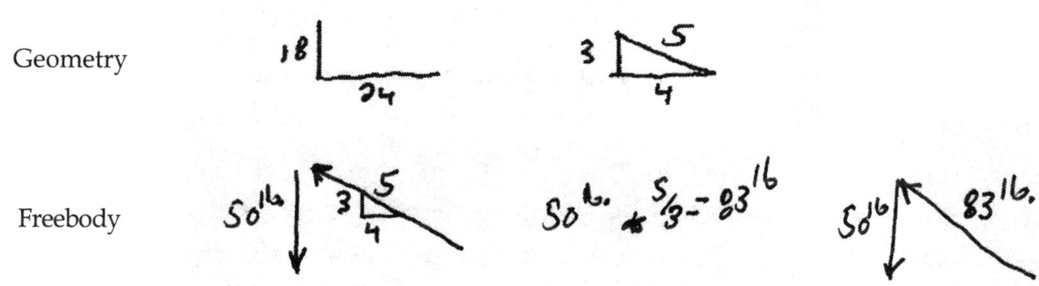

Figure NB-20 Freebody diagram of forces in the Ranum High School Newspaper Bridge

Truss Fun

Chapter 6 also introduced the use of the Finite Element Method as an analysis tool ideally suited for bridge analysis, and provided a program for this use. Figure NB-21 shows the Bollman Tech. High School Bridge organized for entry into the Chapter 6 Finite Element code. Nodes are numbered with larger numbers at intersections, elements are numbered at the member midpoint.

Materials

	Elements	Area	E	
1	1 - 6	1	100000	
2	7,8	1000	1000000	
3	9 -14	0.5	100000	
4	15-20	2	100000	
5	21-24	0.1	100000	
6	------	0.1	100	Elements added for stable solution - will not carry any load due to low E

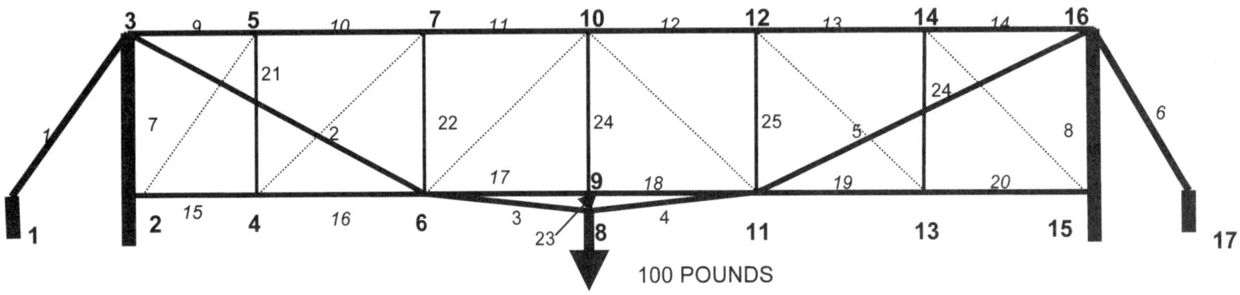

Figure NB-21 Finite Element representation of the Bollman High School Newspaper Bridge

Shown in Figure NB-21 are imaginery members (shown with dashes) which were used in the model to accomodate stability in the analysis. Note that these form triangles with all panels, the program assumes pins at all joints and at joints like 5,7, 10 etc the program would note a zero diagonal term in the solution, or an indication that the structure is not stable. The structure is of course stable, but working joint by joint in the solution, the program can't calculate this. By adding the extra members with low area and very low Modulus of Elasticity values, the solution is stable and the imaginery members are not assigned any load because of their properties.

Forces in members for a 100 pound load at the center of the bridge are shown in Figure NB-22 from the analysis of the bridge. From this figure it is easily observed that the largest forces are in the suspension cable made from rolled newspaper. Forces are shown for elements 1,2, and 3 and range from 30 pounds to 175 pounds for the 100 pound load based on this model of the bridge. Note that the bridge does have 2 sides, 100 pounds on one side would be a total load of 200 pounds in the bridge center, as described before. The 100 pound load was chosen as a convenient load from which other loads could be proportioned. As would be expected, the top of the bridge shows compression, while the bottom of the bridge shows tension (typical of a beam in bending). Forces are the same on both sides of the truss (forces are symmetrical).

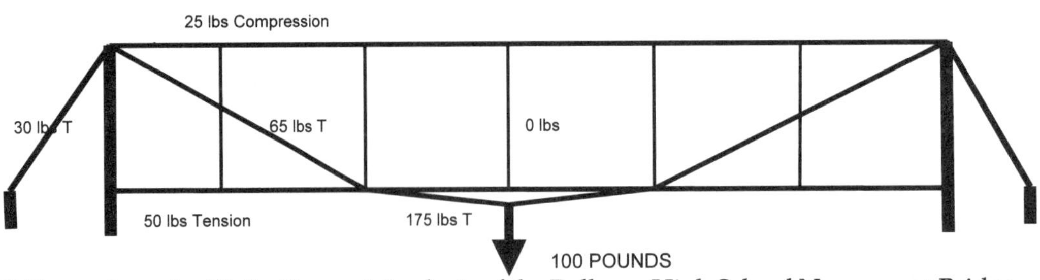

Figure NB 22 Forces from the Finite Element Analysis of the Bollman High School Newspaper Bridge

Design of newspaper bridges is challenging, but fun. It provides a way to practice all principles from this book. Have fun! Be safe!

Index

A
Accident, 92, 94, 105, 131
B
Bending, 58, 59, 85, 88, 89, 92, 103
, 124, 126, 128, 132
Bollman, 16, 18, 21, 25, 26, 27, 28, 29
Buckling, 92, 121
C
Channel, 40, 41, 58, 59, 62, 93
Chord, 14, 15, 19, 28, 30, 40, 41, 64, 114
Compression, 14, 15, 16, 18, 19, 21, 25, 71, 72, 86, 91, 92, 104, 118, 119, 125, 136
Cooper, 22, 23, 30, 113, 114
Covered, 15, 139
D
Deck, 14, 15, 30, 93, 109, 119, 127
Design, 4, 5, 12, 14, 15, 16, 21, 22, 23, 24, 25, 30, 36, 55, 58, 59, 88, 91, 92, 93, 94, 98, 103, 112, 121, 124, 125, 126, 127, 130
Determinant, 16, 70, 72
Directional Cosines, 136
E
E loading, 22, 23, 113, 114, 136
Equilibrium, 67, 68, 69, 70, 71, 72
F
Factor of Safety, 94, 124, 125, 131
Finite Element, 74, 75, 79, 91, 115, 117, 118, 131, 132
Fink, 16, 18, 21, 47
Floor, 15, 21, 24, 25, 27, 30, 58, 93, 94
G
Girder, 12, 47, 48, 50, 52, 93, 94
H
H section, 58, 62
Howe, 15, 18, 36, 37, 38, 59, 64, 90, 93, 98, 109, 114, 124, 125, 126, 130, 135
I
I section, 58
Iron, 5, 12, 16, 25, 28, 40, 41, 47, 48, 90, 93, 94, 125
J
Joint, 12, 14, 16, 21, 36, 40, 41, 91, 110, 121, 124, 125, 126, 127, 128, 130
K
K truss, 16
King Post, 14, 15, 16, 18, 21
Kip, 22, 23, 24, 30, 69, 70, 111, 114, 132, 136
L
L section, 58
Lattice truss, 15, 59, 93, 94
Load, 12, 14, 15, 21, 22, 23, 24, 30, 48, 53, 58, 68, 69, 71, 72, 74, 75, 76, 77, 78, 79, 80, 82, 84, 85, 86, 87, 88, 89, 91, 92, 94, 95, 96, 97, 98, 102, 103, 104, 110, 114, 118, 119, 121, 122, 124, 125, 128, 130, 131, 132, 136
M
Moment, 21, 68, 69, 70, 71, 88, 89, 91, 119, 121, 132, 136
Moment of Inertia, 59, 89, 92, 132
P
Pin, 16, 17, 21, 25, 39, 40, 41, 68, 70, 71, 74, 75, 91, 92, 93, 114, 136
, 15, 16, 19, 21, 40, 41, 68, 69, 70, 71, 73, 104
Pratt, 15, 16, 18, 19, 39, 109, 135
S
S section, 62
Scale
G, 122
#1, 56
1:10, 55
1/1, 55
1/2", 55
1/4, 55
G, 55, 56, 60, 61, 62, 99, 101, 113, 122
General, 55, 56, 57, 59, 60, 61, 62, 64, 65, 95, 96, 97, 98, 99, 100, 101, 102, 104, 113, 122, 124, 125, 127, 136, 145
Graph, 87
HO, 55, 57, 60, 61, 62, 64, 65, 113, 122
N, 55, 57, 60, 61, 62, 101, 113, 122
O, 55, 56, 60, 61, 62
S, 55, 57, 101
TT, 57
Shear, 91
Spring, 72, 74, 85, 87, 90, 97, 103, 113, 114, 132, 136
Stability, 75, 91, 125, 128
Standard
Computer/input, 75, 83
Design(of), 22, 24, 92, 93, 94, 111, 121
General, 24, 53, 58, 59, 115
Materials, 24, 58, 59, 61
standard, 55
Structures, 36, 37, 38, 53, 64, 93, 114, 122

Steel, 5, 6, 12, 16, 40, 41, 48, 58, 59, 62, 63, 64, 74, 87, 94, 95, 97, 99, 100, 104, 105, 106, 119, 126, 127, 128, 138
Stiffness, 16, 72, 74, 79, 80, 81, 82, 84, 87, 91, 95, 96, 97, 98, 99, 101, 132
Strain, 85, 86, 87, 88, 95, 96, 97, 99, 100, 101, 103, 104, 119, 120, 132, 138
Strength, 85, 91, 92, 96, 97, 98, 99, 101, 104, 119, 124, 126, 128, 130, 138
Stress, 30, 68, 78, 79, 84, 85, 86, 87, 88, 89, 95, 96, 97, 98, 99, 100, 103, 104, 118, 119, 120, 124, 132, 138
Strut, 28, 40, 41
Superposition, 89

T
Tension, 14, 15, 16, 18, 19, 21, 25, 59, 70, 71, 72, 86, 91, 104, 118, 121,125,127,136
Through (Truss), 14, 15, 39, 93, 109, 135
Tower , 13, 14, 48, 121
Town, 15
Trestle, 53, 67, 93, 104, 105

V
Vector, 67, 71, 74, 84, 105

W
W section, 58
Waddell, 18 , 41, 44, 45, 46, 105
Warren, 16, 19, 21, 30, 31, 76, 82, 109, 110, 135
Wood, 12, 14, 15, 16, 18, 58, 59, 64, 90, 98, 99, 100, 121, 122, 124, 125, 126, 128, 129, 130

Trusses Old and New
Foreground: Historic Railroad Trusses
Background: Trusses in Denver's New Pepsi Center, Roof and Assembly Cranes, 1999

Items of interest from:

Baha Enterprises, LLC
PO Box 261173
Lakewood, CO 80226

Fungineering - We put the fun in Engineering

Truss Fun Second Edition _____ @ Each 21.95 ____________

_______Please send signed/autographed

Our New CD Learning Product:

The Newspaper Truss

And Other Newspaper Bridges

Fungineering from BaHa Enterprises, LLC

Reference Material and PowerPoint Presentations

Fun Projects in Engineering at High School Level

Using Basic Geometry and Physics

Watch High School Students Take the Challenge

Build and Cross a Bridge From Newspaper

Laugh Learn

Suspension, Truss, Other Types

This CD Contains (over 250 files total):

- Video Clips and Pictures from the First and Later Projects
- Powerpoint Presentations:
  - Introduction to the Project/ View the Results
  - Review of Geometry and Physics for the Project
  - Types of Bridges
  - Analyzing Forces in Bridges Using Simple Geometry
  - Simple Experiments to Find Strength of Bridge Parts
  - Construction Details of The Projects;
  - Tips for Your Own Project ... more
- Cool Web Sites
- Library of Congress References Materials ... more

_____ @ Each 14.95 ____________

Subtotal of Order ____________

Shipping and Handling (Per 3 items) 3.95

CO residents add sales tax ____________

Total Enclosed (Check or Money Order Please) ____________

Ship to:

____
Name

____
Address

____
Address

____
City State Zip Code

____
Country